T0321428

Neurobiology of the Trace Elements

Volume 2

Neurotoxicology and Neuropharmacology

Contemporary Neuroscience

Neurobiology of the Trace Elements, edited by
  Ivor E. Dreosti and Richard M. Smith
  Volume 1: Trace Element Neurobiology and Deficiencies
  Volume 2: Neurotoxicology and Neuropharmacology

Neurobiology of the Trace Elements

Volume 2

Neurotoxicology and Neuropharmacology

Edited by

Ivor E. Dreosti and Richard M. Smith

Foreword by

Sir F. Macfarlane Burnet

Humana Press · Clifton, New Jersey

Library of Congress Cataloging in Publication Data

Main entry under title:

Neurobiology of the trace elements.

   (Contemporary neuroscience)
   Includes bibliographies and indexes.
   Contents: v. 1. Trace element neurobiology and
deficiencies—v. 2. Neurotoxicology and neuropharma-
cology.
    1. Neurobiology—Collected works. 2. Trace elements
in the body—Collected works. I. Dreosti, Ivor E.
II. Smith, Richard M. [DNLM: 1. Trace elements—
Physiology. 2. Trace elements—Deficiency. QU 130
N4938]
QP352.N48   1983   599'.0188   83-8413
ISBN 0-89603-046-6 (v. 1)
ISBN 0-89603-047-4 (v. 2)

©1983 The Humana Press Inc.
Crescent Manor
PO Box 2148
Clifton, NJ 07015

All rights reserved

No part of this book may be reproduced, stored in a retrieval system,
or transmitted in any form or by any means, electronic, mechanical,
photocopying, microfilming, recording, or otherwise without written
permission from the Publisher.

Printed in the United States of America

Foreword

F. Macfarlane Burnet

I have been an interested onlooker for many years at research on the biology of trace elements, particularly in its bearing on the pastoral and agricultural importance of copper, zinc, cobalt, and molybdenum deficiencies in the soil of various parts of Australia. More recently I have developed a rather more specific interest in the role of zinc, particularly in relation to the dominance of zinc metalloenzymes in the processes of DNA replication and repair, and its possible significance for human pathology. One area of special significance is the striking effect of zinc deficiency in the mother in producing congenital abnormalities in the fetus. The fact that several chapters in the present work are concerned with this and other aspects of zinc deficiency is, I fancy, the editors justification for inviting me to write this foreword.

In reading several of the chpaters before publication, my main impression was of the great potential importance of the topic of trace metal biology in both its negative and positive aspects—the effects of deficiency of essential elements and the toxicity of such pollutants of the modern world as lead or mercury mainly as organic compounds. However, with regard to how deficiency or toxicity is expressed in symptoms and pathology, the immense anatomical complexity of the nervous system and the intricate biochemistry of its development and control has much delayed a clear understanding of the functional nature of trace element deficiencies and of the toxic effects of metals.

Another important aspect that emerges is that once a diagnosis of deficiency has been made, it is usually a simple matter to rectify it. I find it impressive that virtually all the clinical types of zinc deficiency, including genetic ones, have been recognized by the positive therapeutic effect of administering zinc salts bv mouth.

When we are dealing with the practical problems of pollution of the environment by small concentrations of mercury, lead, copper, and so on, the main practical difficulty is to establish standard quantitative procedures by which the environmental concentration of the metal in question can be assessed, to decide on the level above which the concentration is not socially tolerable. As with most other problems of environmental pollution, from cigaret smoking to the hazards of nuclear power plants, any decision made is certain to be controversial. From the medical angle the major difficulty is probably the existence of wide diversity among human beings in their genetic susceptibility to the agent and their personal history in relation to the pollutant, which may be extremely difficult to assess since it may involve work and home situations over a decade or more. Equally difficult to quantify are the variations in environmental concentrations of the agent in time and place over the period concerned.

Another very controversial area emerges when a claim is made that a deficiency of an element, for example zinc, during some portion of prenatal and early postnatal existence, can result in retarded mental development. It is hard to envisage any way to overcome the quantitative inadequacies in all such investigations and in practice the number of individuals showing *unequivocal* signs of clinical damage, must be balanced against the cost of doing anything effective to remedy the situation, as modified by the level of education and goodwill in the community and its degree of affluence.

Apart from the social importance of problems arising in the field of essential trace metals and the toxicity of some metals as environmental pollutants, many questions of biological significance await clarification by further research. Several are discussed in chapters of the present volume. One can wholeheartedly agree with Smith and Dreosti's statement that "each trace metal has emerged as essential because it filled a niche in evolutionary progression, and the niches are not related."

From this angle I find special interest in the unique position of zinc, as being characteristically present and essential, as a component of almost all the enzymes concerned with handling the replication and repair of DNA and RNA, as well as in a variety of other enzymes including alkaline phosphatase, carbonic anhydrase, and alcohol dehydrogenase with widely different functions. Recent work in the laboratories of Leslie Angel and Manfred Eiger suggests that the catalytic action of zinc ions may have played a vital role in the prebiotic evolution of polynucleotide precursors of RNA and DNA. It will be a fascinating experience to watch how the wide range of pathological conditions suspected or known to be related to general or organ-localized zinc deficiency will eventually be ascribed to the element's various functions within the body, as a component of metalloenzymes and in other fashions. For the philosophically minded it will be even more interesting to analyze the physical chemistry of the significance of zinc for the catalytic activity of the enzymes containing it, including the ways by which it is incorporated into the enzyme polypeptide during or after synthesis.

Antagonism between trace elements, copper and zinc for example, provides other unsolved problems. Where both are essential to a mammal, it becomes self-evident that a homeostatic equilibrium must be provided for each element. Both in relation to neurobiology and other aspects of physiology, there are strong hints that supplies of trace metals can be directed toward organs or systems where the need is greatest. In pregnancy there is a special cell for zinc in the fetus, and plasma zinc is low in children and adolescents at the periods of maximal growth.

If the current view is correct, that zinc has been essential from the period of prebiotic evolution as part of the mechanism by which the replication of DNA is kept essentially error-free, but yet allows a certain proportion of random errors—around 10^{-8} per nucleotide per replication—it is clear that zinc is intimately concerned with the whole evolutionary process of change by the appearance and eventual expression of mutant genes. Equally, the probably related processes in somatic cells that result in neoplastic growth or the degenerative processes associated with aging seem to have the implication that any foreseeable success either in extending the human life-span or more desirably increasing the proportion of people who can reach 85 with a tolerably good level of health, will require careful consideration of how normal and necessary amounts of zinc and the other essential metals can be maintained in the aging individual.

As a gerontologist I am particularly worried with the extent of mental failure among the old and the increase in intellectually impaired and demented people that must take place in the next few decades. In a recent overview it is stated that around the year 2020 approximately 20% of the population of the USA will be over 65, of whom 10% will have clinically significant mental impairment, half of them sufficiently affected to be labeled demented. The possibility that inadequate zinc or other trace metals may play a part in the breakdown of neuronal integrity is not yet excluded and represents one of the important areas needing further research in the field of the neurobiology of trace metals.

Preface

The science of neurobiology is a relatively modern discipline that has emerged as a coherent entity largely in the last decade. Trace elements too have experienced a recent resurgence of interest, both with the discovery of new essential elements and because of the increasing problem of heavy metal pollution. Not surprisingly, these two expanding disciplines have been found to overlap in a variety of areas and substantial studies have lately been published that describe the extent and complexity of a number of such interactions. Accompanying this burgeoning research interest would seem to be the beginning of a new scientific discipline, possibly best described as "trace element neurobiology."

The multidisciplinary nature of the new speciality has made it necessary to group together workers from widely differing scientific backgrounds including neuroscientists, psychologists, and nutritionists. Research publications in the area cover an extensive field of endeavor and make the collation of relevant information into one comprehensive volume a difficult task. Nevertheless, the current state of the science required that this be done and provided the impetus and *raison d'etre* behind the present text.

In format, Volume 1 has been devoted to chapters relating to deficiencies of the essential trace elements; Volume 2 to the toxic trace elements and to trace elments of neuropharmacological interest, which at present includes only lithium, although it is highly probable that the list in this area will grow with the acquisition of knowledge.

With respect to the essential trace elements, it should be mentioned that, of the 103 elements listed in the Periodic Table, only twenty-five are considered to be required for health and development in animals. Of these, fourteen are called trace elements, which refers to the fact that they occur in the body in "trace" amounts, or more precisely at concentrations less than 0.01% of body weight. Because of their relative scarcity, trace elements are generally involved in a regulatory rather than a structural capacity, but they function through a wide variety of biochemical mechanisms. Indeed, apart from their essentiality and their scarcity, the trace elements share little in common to suggest a collective grouping.

Interest in the toxic metals has growth recently because of their increasing occurrence as environmental contaminants. In no way should they be considered related to the essential trace elements. Nevertheless, because their toxicity becomes manifest at levels of dietary intake or body accumulation that are so low as to be considered "trace" amounts, these two disparate groups of elements have been assembled in this volume under the descriptive heading of "Trace Elements." Lithium, no doubt, falls somewhere within this classification, being a non-essential element, but exhibiting therapeutic properties in certain instances when administered in trace amounts.

In keeping with emerging trends in nutritional and teratological research, the present text highlights the sensitivity of the nervous system to nutritional deficits and to toxic insults that are often manifest at this level before they become detectable in other systems. Also, many contributors have accented the special vulnerability of the fetus to environmental distortions, with the result that much of the discussion of the consequences of elemental deficits or toxicities concerns neural teratogenesis, as well as the pathology of degenerative lesions seen in older animals.

It is our hope that this book will provide a comprehensive discussion of contemporary knowledge concerning trace elements and neurobiology. We acknowledge with pleasure the participation of our distinguished contributors and the initiative and support of The Humana Press. We trust that this volume will, for some time at least, provide a vade mecum for those scientists working in, or entering, this exciting new area of biological research.

Contributors

Dennis P. Alfano, Department of Psychology, and Division of Life Sciences, University of Toronto, West Hill, Canada

Bjorn Arvidson, Neuropathological Laboratory, Department of Pathology, University of Uppsala, Sweden

Satya V. Chandra, Industrial Toxicology Research Centre, Lucknow, India

B. H. Choi, Division of Neuropathology, Department of Pathology, University of California, Irvine College of Medicine, Irvine, California

P. McConnell, Department of Human Anatomy, University of Oxford, Oxford, England

Ted L. Petit, Department of Psychology and Division of Life Sciences, University of Toronto, West Hill, Ontario, Canada

Louis J. Pierro, Department of Animal Genetics, University of Connecticut, Storrs, Connecticut

Michael H. Sheard, Department of Psychiatry, Connecticut Mental Health Center, Yale University School of Medicine, New Haven, Connecticut

Bernard Weiss, Division of Toxicology and Environmental Health Sciences Center, University of Rochester School of Medicine and Dentistry, Rochester, New York

CONTENTS

CHAPTER 1
BEHAVIORAL TOXICOLOGY OF HEAVY METALS
Bernard Weiss

CHAPTER 2
CADMIUM TOXICITY AND NEURAL CELL DAMAGE
Bjorn Arvidson

CHAPTER 3

CADMIUM AND TERATOGENESIS OF THE CENTRAL NERVOUS SYSTEM
Louis J. Pierro

CHAPTER 4

NEUROBIOLOGICAL AND BEHAVIORAL EFFECTS OF LEAD
Ted L. Petit and Dennis P. Alfano

CHAPTER 5
NEUROTOXIC EFFECTS OF LEAD
P. McConnell

CHAPTER 6
NEUROLOGICAL CONSEQUENCES OF MANGANESE IMBALANCE
Satya V. Chandra

CHAPTER 7
MERCURY AND ABNORMAL DEVELOPMENT OF THE FETAL BRAIN
B. H. Choi

CHAPTER 8
ALUMINUM NEUROBEHAVIORAL TOXICOLOGY
Ted L. Petit

CHAPTER 9
LITHIUM EFFECTS ON BIPOLAR (MANIC–DEPRESSIVE) ILLNESS AND OTHER BEHAVIOR
Michael H. Sheard

TRACE ELEMENTS IN NEUROBIOLOGY—CONCLUSIONS
by R. M. Smith and I. E. Dreosti

Contents of Volume 1

Trace Element Neurobiology and Deficiencies

Chapter 1

Behavioral Toxicology of Heavy Metals

Bernard Weiss

1. Introduction

Toxicology is inherently a multivariate discipline, and metal toxicology may be the best example of this characteristic. Metals are so abundant in the environment that some were bound to be deployed in the evolution of biological systems while others took on the roles of competitors or antagonists. Behavior shares this multivariate character with toxicology. Behavior, in fact, is a unique functional measure precisely because it integrates and reflects the total functional status of all of the organism's biological systems, even its compensatory mechanisms. Its advantage as a measure stems not only from this feature, but also from its ability to offer a nondestructive index of functional capacity that permits us to trace the time course of a toxic process.

Behavioral indications of toxicity are common attributes of excessive heavy metal exposure. Table 1, updated from an earlier review,[82] lists some of the adverse neurobehavioral actions of various metals. The cogency of documentation is quite variable, and in certain instances, somewhat flimsy. The more reliable entries are founded on experimental data. Others originated in

Bernard Weiss: Division of Toxicology and Environmental Health Sciences Center, University of Rochester School of Medicine and Dentistry, Rochester NY.

1

Table 1
Neurotoxicity of Selected Metals

Element	Form	Behaviorally significant or detectable effects	
		Acute or subchronic	Chronic
Aluminum	Inorganic		Encephalopathy (dialysis), with speech disorders, aphasia, dementia
Arsenic	Inorganic	Giddiness, headache, extreme general weakness	Polyneuritis with burning and tenderness in limbs; difficulty in walking; paresthesias; fatigue; weakness
	Organic	Optic nerve and retinal damage impairing vision	
Bismuth	Organic	Neuropathy; audiogenic seizures	
Boron	Boric acid, sodium borate		CNS depression; loss of appetite
	Boranes	Restlessness, incoordination, tremors, aggression in mice; headache, dizziness, weakness, disorientation and jitteriness in humans	Similar to acute in mice, but exaggerated
	Tetraphenyl	"Rage"; convulsions	
Cadmium	Inorganic and elemental	Lesions of sensory ganglia in young rats; brain lesions and behavioral disturbances in suckling rats	Elevated locomotor activity in rats treated perinatally; anosmia in humans
Lead	Inorganic		Many reported in animal studies, particularly perinatal exposures; behavioral deficits in children with raised but asymptomatic body burdens; nonspecific and neurological symptoms (peripheral neuropathy) in adults

2

Element	Form		
Manganese	Organic	Hallucinations, convulsions, neurological impairment	Psychiatric symptoms, speech disturbances, Parkinson-like phenomena
	Inorganic	In monkeys, sudden movement, lethargy, tremor at high doses	
Mercury	Organic		Sensory disturbances (restricted visual fields, paresthesias), tremors, ataxia
	Elemental		Psychiatric signs ("erethism"), tremors, ataxia
Nickel	Carbonyl	Headache, delirium, giddiness, weakness, insomnia, irritability	
Selenium		Impairment of vision, weakness of limbs in animals ("blind staggers") following subacute exposure	Dizziness, lassitude, depression, fatigue
Tellurium	Elemental and inorganic		Somnolence, loss of appetite in humans; in rats interferes with acquisition of avoidance performance and discriminations
Thallium		Restlessness, ataxia, tremor, convulsive movements in animals; peripheral neuritis, movement disorders, disorientation in humans	Hind leg paralysis in animals; leg pains, "mild cerebral disturbance," incoordination, sleep disturbances in humans
Tin	Organic, esp. triethyl	Generalized limb weakness, progressing to paralysis in animals; headache, vertigo, photophobia, in humans	Weakness in animals; residual effects in humans include headache, diminished visual acuity, areas of anesthesia
Vanadium	Inorganic salt	"Nervous disturbance" in animals and humans, blurred vision, tinnitus, headache, appetite loss	Tremors and "nervous depression", manic depressive psychosis

case reports and clinical observations. I will survey some of these entries in detail, but will be selective, not exhaustive in treating the published literature. My aim is to illustrate the most essential points.

2. Lead

No other metal has stirred as much experimental effort and policy discussion as lead, and hardly any other environmental contaminant has provoked as much behavioral research. Lead's ubiquity in the environment and its recognized toxicity at higher exposure levels and body burdens have stimulated questions about its safety at lower levels not associated with overt clinical toxicity. The debate centers on behavior and incipient impairment of function. Although children became the focal issue because they were more easily recognized as potential victims, occupational exposures and their consequences are active issues as well because of the swelling documentation that previous estimates of safe body burdens in adults may have grossly underestimated the hazard.

Permeating the debate about behavioral impairment is the issue of body burden measurement. One facet of this issue is the source of the lead, especially the relative contributions from food, paint, or airborne sources such as gasoline combustion. An allied question is how to measure exposure. Most of the literature has relied on concentrations in blood. The deficiencies of this measure, even when reliably performed, are widely acknowledged. Blood lead provides an indication only of recent exposure. If current deficits, say, in test performance, are the result of excessive lead exposure in the past, correlations with blood concentrations are certain to be flawed. For this reason, many investigators have turned to measures, such as concentrations in teeth, providing more accurate estimates of previous lead exposure. Entwined with these questions is one vigorously introduced by Patterson.[56] Since the environment supporting most of human evolution was practically free of biologically available lead, can the body burdens measured in modern humans, which are thousands of times greater, be considered nontoxic even though no significant health hazards are documented? Where could one find an adequate control population for such a survey? The cogency of this question is highlighted by a recent review of trace element requirements in nutrition.[49] It remarked on how long it took to discover the essential role of certain trace metals because purified laboratory environments were required to do so.

2.1. Human Toxicity

2.1.1. TOXICITY IN CHILDREN

If any publication can be said to mark the beginning of our current posture toward the implications of elevated lead exposure in children, it probably would be the one by Byers and Lord.[14] Until then, children who recovered from episodes of clinical lead poisoning were believed to suffer no lasting consequences. By tracing the subsequent history of a group of such children, Byers and Lord uncovered a disturbing spectrum of behavioral and neurological problems. The children performed poorly in school, and, in addition displayed conduct disturbances such as hyperactivity. Although, by the standards of modern clinical pharmacology, the study suffered from many flaws, it soon inspired a series of additional studies aimed at the question of enduring consequences. These studies, too, were flawed by the absence of adequate controls, by lack of body burden measures, and by crude indices of behavioral function. Propelled by consistencies difficult to dismiss, and by the emergence of the new discipline of behavioral toxicology, the issues began to converge toward a different question: the relationship between raised body burdens in asymptomatic children and measures of behavior.

Most of these later studies relied upon standardized psychological tests as criteria of performance. To appreciate the meaning of the data such tests yield, it is essential to understand the process by which they are constructed. Tests are samples of behavior. They are developed as selection tools or to quantify an individual's standing in a population on some property such as achievement in arithmetic. Psychometrics strives for two features in a test. One is reliability. That term describes the stability of scores, that is, the reproducibility of performance from one test occasion to another. Reliability is expressed as a correlation coefficient; a test with low reliability provides scores with low correlations between one administration and another, so it yields undependable data. Some of the instruments and measures chosen for evaluation in the lead literature are of low or even uncalculated reliability. The other feature is validity, or the degree to which the test measures what it purports to measure. Validity is determined in different ways. Predictive validity is sought for tests such as the College Boards, which aim for high correlations with future grades. Content validity is an empirical judgment, consisting essentially of examining the test items. Fluency in a foreign language, for example, might be tested by requiring translation of a specified passage in that language. A test may possess construct validity, a kind of conceptual validity determined by the congruence of its contents with the property to be measured. Such an approach to validity is typically taken with intelligence tests for children. Test framers assume that as children develop they become more and more capable and informed. A child able to define vocabulary items with about the same facility as other children its age would receive a score close to normal. A child with a more extensive vocabulary would score above normal. If intelligence test scores are transformed into equivalent age performance scores, a quotient (IQ) can be calculated from performance age as the numerator and chronological age as the denominator. An IQ of 100, then, would represent test scores prevailing for that chronological age, and an IQ above 100 would represent performance beyond what would be expected for that age.

Defining "normal" is the problem. The test designer must choose a sample of children on whom to standardize the test. But a child with a background different from those on whom the test was standardized cannot be labeled with a score that actually describes its "intelligence." A test standardized on white, middle class children, like the Stanford-Binet, will give misinformation about a sample of black, lower class children, although it might help predict, because of its academic orientation, performance in the classroom. Since so much of the lead literature deals with children rather different from those on whom many of the tests have been standardized, reviewers as well as investigators should be aware of the discrepancies and their potential distortion of the outcome of a study. They might, for example, greatly inflate variability, and reduce the sensitivity of a comparison. Or, they might simply be unsuitable measures of competence in the sample chosen.

Unreliability and lack of validity will attenuate test score differences between groups selected on the basis of a biological measure such as blood lead. An unreliable test introduces a large random element that reduces the proportion of total variance accounted for by exposure differences. And, if the test itself does not measure what it purports to measure (such as, "intelligence"), it may reflect some possibly irrelevant qualities associated with a particular ethnic culture or social class. For this reason, some recent studies have more carefully investigated parental and class contributions.

Perino and Ernhart[58] tried to control for possible spurious relationships by performing a multiple regression analysis in which age, parent IQ, and birth weight were treated as covariates; that is, their influence on calculations of statistical significance was corrected for. The authors compared black preschool children whose blood lead concentrations fell either into the range of 40–70 μg/dL or into the range below 30 μg/dL. The test instrument was the McCarthy Scales of Children's Ability, a test battery specifically designed for assessment of younger children and containing several motor control components. The multiple regression analysis demonstrated a significant correlation between lead body burdens and scores on the verbal, perceptual, and cognitive subscales of the test battery. An ancillary finding captured considerable interest. The correlation coefficient between maternal and child IQ reached 0.52, a typical value, in the lower lead group, but only 0.10 in the higher lead group. One explanation for the difference, tentatively advanced by the authors, is that the lower lead levels in the controls arose from more careful parental monitoring. Without better exposure histories, however, the data remain puzzling.

The problems arising from trying to estimate exposure from blood lead values stimulated Needleman et al.[55] to seek another approach to the question of low-level lead toxicity. In what is acknowledged to be the most definitive study so far, these investigators classified lead exposure on the basis of dentine concentrations in shed deciduous teeth collected from over 3000 first and second grade children in the Boston area. After excluding children for various reasons that would confound the central question, 58 children with high and 100 with low dentine levels were compared by a battery of psychological tests, including an intelligence test, other cognitive tests, tests of auditory and language processing, reaction time, and tests of motor coordination. To ensure that the groups, except for tooth lead, were comparable, they were examined on 39 nonlead covariates. Those few that differentiated the groups, such as maternal IQ, were entered as covariates into an analysis of covariance so that they would not inadvertently contribute to the evaluation of observed differences in performance. Children in the high lead group differed from the children in the low lead group on five of the 14 subtests of the intelligence test and on full-scale IQ, on reaction time, on measures of auditory processing, and on several other indices. An earlier screening of these children, 4–5 yr previously, showed mean blood values of 35.5 μg/dL in the high and 23.8 μg/dL in the low dentine lead group. This is a surprisingly small difference to account for about a 4-point difference in IQ.

For toxicology, perhaps the most intriguing outcome of the study was the function obtained from a teacher rating scale of various classroom behaviors for 2146 of the children. Classifying these children into six lead concentration groups, and plotting their scores on the rating scale, yielded a clear concentration–response function for most of the 11 items, suggesting that a "threshold" or "no-effect level" may be impossible to calculate. This plot would be even more striking if the indices had been combined into a single score, and the data even more convincing if subjected to multivariate analysis. A parallel study in Germany by Hrdina and Winneke, but not yet fully reported and with a much smaller sample than that of Needleman et al., found results much like the Boston study.

The report of a British government commission on lead[25] expressed several reservations about Needleman et al., but misinterpreted two important aspects of that study. First, although the teacher's rating scale was criticized as not a standardized instrument, it was not used to classify children, only to prompt a response from the teacher. The striking dose–response functions, in fact, could be considered the validation of the inventory. Second, the report pays scant attention to a pervasive finding in toxicology, namely, the enormous variability in sensitivity within a population.

Some facets of these problems were addressed by Yule et al.,[86] originally members of the group preparing the British report. They were particularly anxious to procure further information about the effects on children of blood lead concentrations below 35 µg/dL. Their study population consisted of 166 children 6–12-yr-old. Only one of the children gave a blood value above 30 µg/dL (32 µg/dL). The children were given a battery of tests that included both achievement and intelligence measures. The statistical analyses were quite extensive, relying on multivariate methods to partial out the contribution of variables such as social class. Even with social class statistically nullified, an average difference of 7 IQ points emerged between children whose blood lead values fell below 12 µg/dL and those whose levels lay above 13 µg/dL. In addition, significant correlations emerged between blood lead concentration and measures of IQ and educational attainment (except for mathematics) even after partialing out age, sex, and social class.

The blood lead values in Yule et al. were based on bloods drawn 9–12 months before psychological testing. The authors remark as follows on the significance of that interval[86]:

    If blood lead levels are so labile as some authorities be-
lieve, the finding of small, statistically significant relationships with attainment and intelligence are all the more
surprising.

    This comment highlights issues that are not discussed often
enough.

2.1.1.1. The Accuracy and Meaning of Blood Lead Esti-
mates.   Even after the far from trivial problems of precision and
reproducibility have been overcome, what information is really
conveyed by single, isolated blood lead concentrations? Given the
lability of such values, can any reasonable inferences be derived
about exposure history? Overt poisoning is most frequent among
children 1–3-yr-old, but most studies have been performed on
older children. Even an interval as short as 1 yr can introduce confusion. Baloh et al.[6] found five children out of a small sample
whose assignment to a "low-lead" group (less than 30 µg/dL)
turned out to be invalid because the later assay showed them to
have elevations above that limit. Dentine lead provides an inte-
grated or cumulative estimate of exposure, but not the distribution
over time and possible sharp peaks. Rutter[64] observed:

> All these considerations mean that the differentiation
> of high-lead and low-lead exposure groups (especially on
> single blood estimations) will be somewhat unreliable,
> therefore the differences in outcome between them are
> likely to constitute an underestimate of the true differ-
> ences between populations with chronically increased
> and acceptably low levels of lead exposure.

2.1.1.2. The Distribution of Lead Tissue Levels.   Like many such
measures, the distributions of blood and dentine lead describe a
lognormal or logistic function and not the symmetric normal
curve on which most parametric statistical tests are based. In the
study by Yule et al., the mean value was 13.52 µg/dL (SD = 4.13),
but, because of the skewed, approximately lognormal distribution
of values, natural logarithm transformations were used in the statistical analysis. Dentine leads in Needleman et al. also resembled
such a distribution, that is, one with most values concentrated in a
relatively narrow range, but with a long tail. (Income distribution is
typically best described by the lognormal distribution).

    The features of such a distribution can be visualized by Figs. 1
and 2. Figure 1A is a histogram based on the 166 subjects from Yule
et al. Observe the skewness. Figure 1B is a distribution generated

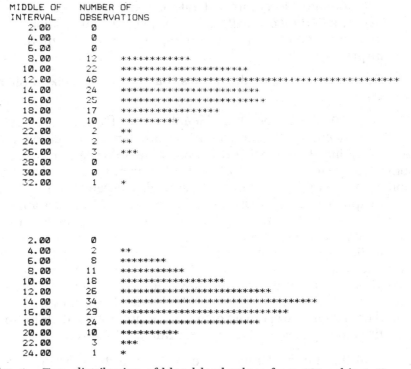

FIG. 1. Top, distribution of blood lead values from 166 subjects.[86] Bottom, values drawn randomly from hypothetical normal distrubition with same mean and SD as A.

from random numbers distributed normally with the same mean and SD as the original data. Note the symmetry. Another view of the contrast is plotted in Fig. 2, where Fig. 2A represents the cumulative distribution function based on Yule et al. and Fig. 2B represents the synthesized data. Yule et al. so far is the only paper in the human literature explicitly to take account of this property of the data in statistical calculations. But the implications may be critical to the debate about risk. Do such distributions indicate that susceptibility to lead toxicity is not normally distributed? That, instead, a small proportion of children account for differences between groups? That the 4 point IQ difference in the Needleman et al. study may be the contribution of 25% of the children who provide, so to speak, 16 IQ points each to the total?

That the data from the many published studies are inconsistent and often puzzling is no surprise. Lead is only one of the hazards imposed by the environment and its influence can be overshadowed by poverty, malnutrition, parental deficits, and many

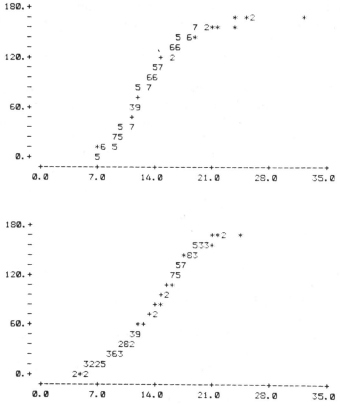

Fɪɢ. 2. Top, cumulative distribution (blood lead vs frequency) of data in Fig. 1 (top)[86]; bottom, cumulative distribution function based on random sample depicted in Fig. 1 (bottom).

other variables and events. But the sum of the published literature, especially the more recent and better-designed studies, grants it a significant role.

2.1.2. Lᴇᴀᴅ Tᴏxɪᴄɪᴛʏ Iɴ Aᴅᴜʟᴛs

Largely on the basis of data from neuropsychological testing, although other criteria have been included in the decision, the body burdens recognized as hazardous to adults also have fallen consistently during the past few years. As in children, overt clinical toxicity was not the issue. Incipient, subtle impairment displaced gross deficits as the basis for occupational standards. Yet, even before the discipline of behavioral toxicology attained formal recognition, industrial hygienists and physicians were alert to the potential signs of early lead exposure. Some warned of personality changes, such as irritability, as precursors of more severe reactions or as simmering reflections of borderline clinical toxicity. Several investigations of lead-exposed workers have now been published that document impairment in groups with moderate body burdens and with no evidence of clinically detectable deficits or illness.

Hanninen and her coworkers have pioneered in the application of psychological testing to occupational health assessment.[35,36] They developed a test battery, individually administered, that includes elements of adult intelligence tests, memory assessment, and perceptual–motor performance. They also have made use of questionnaires to probe for subjective correlates of exposure. Only lead workers whose occupational history had been closely monitored by blood lead samples, and whose blood concentrations had never exceeded 70 μg/dL were taken into the study. The most convincing differences between lead-exposed and control workers[35] were contributed by two subtests, block design and digit span, from the Wechsler Adult Intelligence Scale (WAIS), and by a test of the ability of the subject to reproduce a drawing. The later publication[36] compared two groups of lead-exposed workers with only moderately elevated blood leads to a control group, this time with the aim of determining their responses to an extensive questionnaire embracing a wide range of subjective complaints. Some of these were mainly psychological (for instance, depression) and others were mainly somatic (for instance, headaches). The incidence of complaints was significantly higher in the two lead-exposed groups. Hanninen et al. did not provide the distribution of responses within the groups, unfortunately. Given the pronounced variation in susceptibility among individuals, I would like to see if the group differences arise from a few individuals who skew the distribution.

Although the kinds of subjective complaints assayed by Hanninen et al. are compatible with the observations advanced earlier by occupational health specialists, they may not be directly related to central nervous system dysfunction. Two groups of investigators sought more specifically to evaluate central nervous system function by applying tests and techniques found useful in assessing deficits arising from brain damage. Valciukas et al.[74] sought relationships between cognitive performance and exposure indices in a group of workers from lead battery plants. Reasoning that the lability of blood lead concentrations would thwart such an aim, they turned instead to measurement of zinc protoporphyrin (ZPP) IX. At the step in heme synthesis at which iron is incorporated into heme, lead induces zinc incorporation instead. ZPP then corresponds to the lifespan of the red cell, 120 d.

Valciukas et al. chose three performance assays as dependent variables. Block designs, taken from the WAIS, require that specified patterns, printed on cards, be matched by manipulating a set of blocks segmented diagonally into red and white sections. The digit–symbol substitution test is a test of coding ability. Numbers from 0 to 9 are keyed to correspond to a set of abstract symbols. The subject is required to write in the appropriate number under each symbol. Score is calculated as the number of entries within a specified time. The embedded figures test originally was devised to assess deficits resulting from penetrating head wounds and subsequent brain damage. A confusing overlay of line drawings is presented to the patient, who is asked to distinguish the various objects represented. With these three tests, Valciukas et al. were able to demonstrate a significant relationship of ZPP with blood lead. Blood lead values were not significantly correlated with performance.

Assessment of cognitive performance was also the path chosen by Grandjean and Beckmann,[34] who compared such data with results of a subjective symptom questionnaire administered to the same workers. With a battery of tests, including those applied by Valciukas et al., they demonstrated a significant relationship between performance and ZPP. Blood lead concentration might have offered an informative counterpoint; perhaps Hanninen et al.[36] found a significant relationship with subjective indices because blood lead reflects some features of prevailing health status independent of longer term consequences for central nervous function.

2.2. Implications

Gradual lowering of lead body burden estimates believed to be associated with impaired health, especially neurobehavioral function, has marked the literature during the past decade. Although some of this trend might be attributed to the recognized tendency to publish only positive results, it more likely results from improvements in experimental design and refinements in relevant measures of both lead exposure and of behavior. But it can hardly be argued that the full panoply of possible techniques has been exploited. For example, the discipline of applied behavior analysis, which grew out of the need to measure the consequences of behavior modification procedures, provides an elegant technology for quantitative assessment of behavior in the natural environment, a technology enriched now by the availability of unobtrusive, portable digital recording devices. Since so much of the literature of applied behavior analysis is based on classroom observations, and since so much of the child's functional deficits first become obvious in that structured and demanding environment, behavioral recording in the classroom seems an obvious supplement to, if not substitution for, assays such as intelligence testing. Only the Needleman group, however, seems to be promoting any efforts in this direction (personal communication).

Rigorous assessments of sensory and motor function are also absent. Conventional clinical techniques are unlikely to detect the subtle deficits said to arise from moderately elevated lead exposures. Visual science, for example, now provides assessment tools, based on engineering models of the visual system, capable of probing the boundaries of visual function.[47] Motor function also demands more refined procedures than the industrial screening tests or measures of motor conduction velocity used by some investigators. Evaluation of fine motor control and perceptual–motor performance requires an engineering analysis similar to those applied to studies of Parkinsonism.[1]

2.3. Animal Studies

2.3.1. PERINATAL EXPOSURES

Lead experiments on animals were discouraging efforts before 1966 because it seemed impossible to devise an appropriate animal model, that is, one reflecting the pattern of adverse effects observed in humans. The model devised by Pentschew and Garro[57] provided a novel method for producing lead toxicity in animals and evoked a resurgent interest in the problem. Their contribution was to use the nursing rat dam as the source of lead, a maneuver that exposed the developing pup at a time when circumstances conspired to make it exceedingly vulnerable. The immature brain, immature gut, and milk diet were three of these factors. Feeding a diet of 5% lead carbonate to the dam produced pups that developed with stunted and retarded growth, ataxia progressing to paralysis, and marked central nervous sytem lesions.

The behavioral accompaniments of neonatal lead exposure attracted interest after reports of marked elevations in motor activity in the offspring of dams administered lead in drinking water during the period of nursing.[65,67] Interest was amplified when the phenomenon was confirmed in rats whose mothers had been fed a diet containing 4% lead carbonate. Since then a large literature has blossomed, recently reviewed in detail.[13] It is a complex literature to review because the methodology of activity measurement is not trivial; different measures of activity are not highly correlated with each other and are greatly dependent on specific features of the apparatus and definition of the response. Problems of interpretation also arise when biological indices of exposure are lacking, when maternal food reduction and retarded neonatal growth ensue, and when high doses induce overt toxicity. Furthermore, the simplistic assumption of congruence between the childhood syndrome of "hyperactivity" and elevated motor activity that stimulated much of this work is not warranted. In fact, the new diagnostic manual of the American Psychiatric Association has conferred a different label on the syndrome: Attention Deficit Disorder. A substantial literature on learning and its modification by lead exposure also has accumulated. Again, the situations chosen by individual experimenters are so disparate, and the treatments so numerous, that few generalizations can be extracted except that, at certain doses, in certain situations, lead treatment, especially to neonates, impairs learning ability.[13]

Rather than discuss these aspects of the behavioral toxicology of lead, which have been reviewed very thoroughly, I will focus on an aspect that offers more potential than so far has been exploited. Operant, or schedule-controlled behavior, has enjoyed a prominent role in behavioral pharmacology since that discipline's beginnings in the 1950s. It achieved such a role because it offered a technology compatible with the questions that pharmacology needed to ask about drugs acting on the central nervous system; that is, what behavioral mechanisms or functions are altered by specific drug treatments, and how can precise dose–effect functions be established for behavioral endpoints? Although, when posed in a toxicologic context, these questions are more complicated because of the problem of irreversibility and the frequent necessity to study chronic treatment, they address the same principles.

Schedule-controlled behavior is associated with the name of B.F. Skinner, because of his compelling demonstrations that different patterns of behavior are maintained by arranging different relationships or contingencies between a behavioral act and its effects. If food is delivered to a hungry animal for every 50 responses on a response device such as a lever, it emits a different pattern of responses in time than if food regularly follows a response about every 5 min. The ability, in the same situation, to generate different patterns of behavior that were differentially responsive to drugs provided behavioral pharmacologists with an enormously flexible set of tools. Generality among species also conferred a great advantage. Finally, reproducibility from laboratory to laboratory was remarkably high, despite many variations in apparatus design, protocols, and other aspects of the laboratory setting. Together with the stable, quantifiable baselines that operant behavior provided, these advantages made operant techniques dominant choices when the most critical questions about drugs and behavior arose.

One of the earliest contributions to the toxicology literature came from Van Gelder et al.[75] who, because they were housed in a school of veterinary medicine, turned to sheep as an experimental subject. They fed adult females 0, 120, or 230 mg of lead daily. The animals were trained to press down on a large pedal to obtain deliveries of corn. Such deliveries were governed by what is termed a Fixed Interval (FI) 30 s schedule of reinforcement. That is, the first response after the end of a 30 s interval triggered delivery of food. The start of each interval was marked by the preceding food delivery. Responses during the interval had no programmed consequences, but were counted and analyzed. The absence of consequences allows the number and distribution of responses within the interval to vary widely without affecting the regularity of food delivery, so that both features of performance can serve as indices of drug or toxic actions. The usefulness of such measures has been demonstrated many times in behavioral pharmacology. The sheep in this experiment worked only 15 min daily (about 30 intervals) and were trained for 7 weeks to stable performance before lead treatment began. Total daily responses of the lower dose group exceeded that of the controls, but those of the higher dose group were decreased. The difference between the two dose groups was statistically significant. Had more individual data been presented, it would have been possible to search for the source of this difference and the lack of difference with controls. Were there, as I would expect, wide individual differences that made conventional statistical tests inappropriate?

Rats were the subject of a study by Zenick et al.[87] The test animals were born to mothers who had been exposed daily to 750 mg/kg of lead acetate delivered through their drinking water, consuming it before mating and during pregnancy and lactation. Once weaned, litters from half of the lead-treated dams continued on that regimen, and litters from the other half continued on tap water. Control litters were never exposed to lead in drinking water. Such an experimental design confounds litter treatment; it would have made more sense to divide the individual litters. Training began at 42–49 d of age. The rats were reduced to 80% of their free feeding weight and maintained at a corresponding level, relative to free-fed littermates, during behavioral testing. They were tested in experimental chambers containing a lever connected to a switch that served as the response device, and a feeder that delivered 45 mg food pellets when reinforcement was programmed. After the rats had learned to depress the lever to obtain the food pellets, a fixed interval 1 min reinforcement schedule was imposed. What emerged was a markedly lower frequency of reinforcement (completed intervals) on the part of those animals treated both before and after weaning. The pattern of responding during intervals also was analyzed, with no differences among groups apparent, but the methods of analysis are unclear and not congruent with those refined by operant experimenters. No biological measures were presented, and the treatment apparently produced decreased birth weights, so that most aspects of the experiment remain a puzzle.

Some of the puzzling aspects of the report by Zenick et al. were resolved by a more recent experiment.[2] Four groups of rats were constituted. One group served as no-treatment controls (N-N). Group 2 was treated only during nursing by providing the dams with 0.2% lead acetate (1090 ppm Pb) as drinking water (group PB-N). Group 3 received lead in drinking water after weaning (group N-PB). Group 4 was exposed both before and after weaning (group PB-PB). Each group consisted of 10 rats. Behavioral training began on postnatal day 58. Final performance was measured by what is called a multiple reinforcement schedule, that is, alternating individual schedules of reinforcement each designated by a different stimulus. One component of this multiple schedule was a fixed interval (FI) 1-min schedule, identified by a steady light over the response lever. A second component was a fixed ratio (FR) reinforcement schedule, which, as the name implies, required that a specified number of responses be made to secure food delivery. The size of the ratio varied at different times during the experiment. The FR and FI components alternated during the course of each 30-min experimental session, and were separated by periods called time outs (TO) lasting 15 s and during which lever responses had no programmed effects. The FR component was identified by a flashing light above the response lever and the TO by turning off the light in the chamber.

Several different measures of performance were calculated, but the one on which the bulk of the analyses are based is the median interreponse time (IRT). The experiment was controlled by an on-line digital computer that also recorded the intervals between successive responses. Distributions of such IRTs have proven revealing in many behavioral pharmacology experiments[81] and provide insights into many schecule-induced phenomena.[80] The median IRT sometimes is a better index of performance than rate of responding computed simply as number of responses per unit time because it is resistant to distortions introduced by long and variable pauses in responding, and here was recorded with a temporal resolution of 10 ms. Figure 3 shows, for all groups, changes in median IRT during sessions three and five on the FR component of the multiple schedule. The distributions for the FI component were similar. Statistical analyses indicated that the differences suggested by the figure were statistically significant over the first 15 d of training on the multiple schedule. That is, the two groups treated postweaning (N-PB and PB-PB) emitted longer median IRTs than the controls (and produced lower response rates). The group treated only during nursing (PB-N) emitted shorter IRTs than the controls, but the differences did not achieve statistical significance. One reason is the common one of differential susceptibility. Note the clustering of five of the PB-N rats at the low median IRT values. This differential effect, however, was dissipated by the five apparently insensitive rats, so that statistical techniques, with such a small sample of subjects, indicated no reliable effect.

Corresponding blood and brain levels of lead showed pronounced elevations in both of the postweaning groups. On postnatal day 130 the median blood concentrations were 66 and 64 μg/dL for groups N-PB and PB-PB, respectively. Brain leads, measured after sacrifice between postnatal days 141 and 143, came to 375 ng/g for the median of both groups. Both the behavioral and body burden measures indicate that the presumed insensitivity of mature organisms to lead toxicity[13,41] is a issue that needs reexamination.

The most cogent argument for such a reexamination comes from an experiment reported by Cory-Slechta and Thompson.[20] These authors exposed rats after weaning to 0, 50, 300, or 1000 ppm lead acetate in drinking water, then trained them to perform on a fixed-interval 30-s schedule of food reinforcement. Both the 50 and 300 ppm groups (group size in the experiment varied from 4 to 6 animals) showed markedly elevated rates of responding compared to controls. These rates gradually declined to control levels. Rats consuming 1000 ppm showed lowered rates at first that later drifted upwards to control levels and beyond.

These data also present a puzzle in the form of what appears almost to be an inverse dose–effect function. The most prolonged and consistent rise in rates appeared in the rats treated with the lowest concentration. A transparent hypothesis to account for such an effect would posit a dual action of lead, one promoting

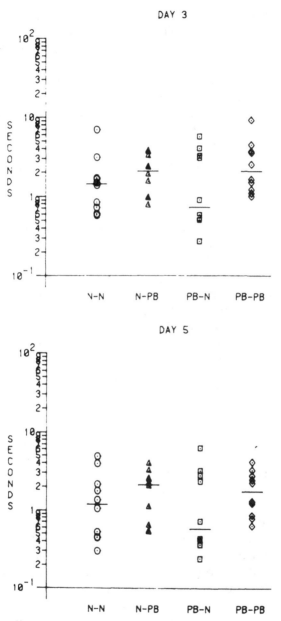

FIG. 3.   Distributions of median interresponse times on FR20 in four lead treatment groups. N-N = controls; N-Pb = postweaning exposure only; Pb-N = preweaning   exposure   only;   Pb-Pb = preweaning   plus postweaning exposure. For preweaning exposures, nursing dams were given 1090 ppm Pb in drinking water. For postweaning exposures, offspring were given the same concentration. Horizontal lines represent group medians.[2]

increased responding, and one, by straightforward toxicity, interfering with responding. The gradual rise in rate displayed by the highest dose group is somewhat consistent with such a mechanism. A systematic replication,[22] which incorporated serial blood lead measures, confirmed the inverse dose–effect function and provided further clues about its source. A group exposed to 500 ppm and observed over a prolonged period of time exhibited a gradual increase in rates on a fixed-interval 1-min schedule, finally exceeding groups treated with 50 and 100 ppm.

Another intriguing operant experiment is particularly notable because it is one of the few studies on primates.[61] The monkeys (Macaca fascicularis) were dosed orally with 500 μg/kg/d of lead (as acetate) from birth. Four animals were assigned to the experimental and four to the control group. Until 200 d of age they were fed a milk formula, then switched to a solid primate diet. The lead-treated animals showed peak blood lead concentrations of 50–69 μg/dL at 200 d of age. These values fell gradually to a stable level of 20–30 μg/dL, in part, at least, because of the change in diet.[50] At the start of the operant study, the monkeys were 2.5–3-yr-old. After being reduced gradually to 85% of their free-feeding weight, they were trained to press a plastic disk to earn a small amount of apple juice. Once this behavior was stable, they were switched to an 8-min FI reinforcement schedule, with the FI components separated by TO periods 90-s long. Rate differences during both the FI and TO appeared, but the most striking distinctions between the control and treated monkeys developed in the pattern of responding. The exposed monkeys tended to respond in bursts, unlike the controls, which tended to respond at a steady rate. This pattern difference also produced a marked difference in the median IRTs of the two groups; those for the treated animals were notably shorter.

The operant (schedule-controlled) behavior studies reviewed here represent selections from a still sparse literature. Although several others have been published, they are not as revealing or as interpretable. Data from pigeon experiments are difficult to interpret because the nucleated erythrocyte of avian species makes blood lead values misleading in mammalian terms. Furthermore, lead treatment in pigeons produces crop stasis, so that although food is ingested[27] and the pigeon's weight seems stable, the food is not being digested.[21]

Although such a small cadre of studies is not a sufficient basis for clear conclusions, they do provide hints for significant questions. They suggest, for example, that sensitivity to lead exposure is magnified during the aquisition of performance.[2,20,21] Such transition states provide special problems in analysis, but digital computer technology makes them easily manageable.[80] The temporal patterning of responses also is amenable to microanalyses, and likely to be revealing.[61] And, as the human investigations demonstrate, close scrutiny of individuals is essential; conventional group statistics may be rather misleading in situations eliciting responses in only a subgroup of the sample under study. Finally, these studies tend to support the contentions of those who assert that the human data indicate adverse effects of lead at tissue levels previously believed to be free of risk. As Rice et al.[61] noted, their juvenile monkeys displayed behavioral aberrations at blood lead concentrations of 20–30 μg/dL, a range of values found in many children, who, like the monkeys, show no clinically detectable problems.

3. Mercury

Like lead, mercury also was mined in antiquity and found many uses in preindustrial civilizations. In certain respects, its toxicity is even better documented, largely because its liquid state inspired widespread medical applications. For hundreds of years it was applied in unctions, usually to patients with syphilis, producing toxic reactions in physician and patient alike. It also was incorporated into anthelmintics, ointments for skin diseases, and even teething powders. For some years, it served a useful role as a diuretic. Mercury also enjoyed wide industrial use. It is still a major component of electrical devices such as meters and an effective fungicide.

Chemical speciation is especially critical when assessing mercury toxicity. Metallic mercury, because of its volatility, presents a hazard that seems to undergo repeated rediscovery. Its ability to seep into fissures in surfaces such as floors, from which it volatilizes, often leads to neglect of necessary precautions because it is not visible. Mercury vapor also reaches the brain far more readily than the ionic forms. Since the conversion of elemental mercury to mercuric ion by blood is a slow process compared to the time required for transport from lung to brain, and since elemental mercury seems to penetrate readily into brain,[15] CNS tissue may contain ten times more mercury after vapor exposure than after a comparable intravenous dose of mercuric salt.[9] The mercury is then retained in brain after oxidation because the

blood–brain barrier hampers its excretion. Some organic mercurials, the alkyl compounds specifically, can be far more toxic. Of
these, methylmercury is the most potent. Although recognized as
extremely toxic after its synthesis in the nineteenth century, it
claimed widespread notice only after identification as the agent
responsible for Minamata disease. Its notoriety ballooned with the
further discoveries that species of marine fish such as swordfish
contained high flesh levels of mercury, and that many freshwater
fish had been contaminated by mercury dumped into bodies of
water. Benthic organisms methylate inorganic forms to produce
the toxic alkyl compound.

3.1. Metallic Mercury

3.1.1. ADULTS

Although mercury mining inspired the first industrial hygiene legislation on record,[38] and epidemiologic surveys of workers have
been conducted since early in this century, systematic investigations are a recent development. A massive study of the hat industry stands as a pioneering effort.[54] Mercuric salts converted the
stiff, straight hairs clipped from animal fur into a limp, flexible
form much easier to mold. The process exposed workers to mercury vapor concentrations in excess of the then prevailing
standard, and engendered many cases of both overt and covert
toxicity.

The most frequently noted sign of mercury intoxication is
tremor. It often was called "hatters' shakes" because so many
workers in the felt hat industry suffered from tremor, or "Danbury
shakes" because Danbury, Connecticut, was the center of the industry. Hatters' shakes usually developed first in the facial
muscles, especially the eyelids, progressed to the fingers, and
finally overcame the entire hand. As exposure continued, tremors
attacked the tongue, speech became thick and slurred, and the
victim began to walk with a jerky, ataxic gait. Since the tremors
worsened when the victim tried to control them, as when grasping
an object or when writing, they were diagnosed as intention tremors.

Neal et al.[54] attempted some rough quantification of the
tremor by having the worker lift a flat wooden strip with a small
block placed on one end, then return it to the table surface. Tremors were rated as slight, moderate, or severe, and the ratings revealed consistent relationships with the number of years employed in the industry and the estimated exposure concentration.

More recent attempts at quantification have taken advantage of electronics and computer technology to extract more meaningful indices from the tremor. Wood et al.[84] recorded tremor from workers exposed to mercury vapor in a factory that used mercury in a process to calibrate pipets. Tremor was recorded with a strain gage, amplified, and fed to the analog–digital inputs of a digital computer. A mathematical technique know as the Fast Fourier Transform decomposed the signal into its constituent pure frequencies, yielding a plot of frequency against power (Fig. 4). The most notable feature of this plot is the multiple modes appearing in the spectrum when the worker was first admitted to the hospital. That is, the tremor, rather than being dominated by a single mode, reflected contributions from several possible sources or an inherent irregularity, either of which springs from disordered control mechanisms. Removal from the source of exposure for 9 months led to a marked fall in plasma mercury levels and a concomitant fall in tremor amplitude (or power) accompanied by a narrowing of the spectrum and the restoration of a single mode.

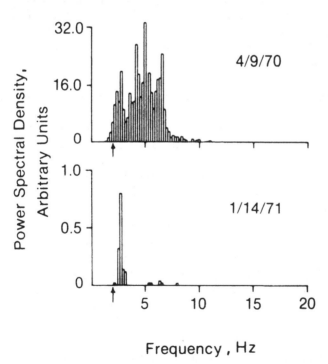

FIG. 4.   Power spectral density plots showing spectra (top) at time of discovery of high plasma mercury and nine months later (bottom) in female patient.[84]

Table 2
Symptoms of Erethism

• Hyperirritability	• Timid, shy
• Blushes easily	• Depressed, despondent
• Labile temperament	• Insomnia, fatigue
• Avoids friends and public places	

This approach to quantification has been adopted to monitor workers in chlor-alkali plants, where huge mercury electrodes are used to process brine into chlorine and caustic soda.[43] Workers with high urinary levels of mercury also show multiple peaks in the power spectrum that fade when the worker is removed from exposure.

Mercury poisoning is also accompanied by a galaxy of psychological symptoms subsumed under the label "erethism" (Greek for "red" or "irritation"). The list of symptoms is given in Table 2. It is easy to see why a clinician unfamiliar with mercury poisoning or who fails to obtain an occupational history might be misled into a diagnosis and treatment regimen based on psychological variables alone. The German chemist, A. Stock, suffered from erethism and many physiological problems during the years he spent at the chemist's bench inhaling mercury vapor. He termed his affliction "micromercurialism," to distinguish it from overt, severe mercury poisoning.

As might be expected, cogent quantification is lacking. One investigation relied on questionnaires completed by chlor-alkali workers.[68] The clearest relationship to emerge showed loss of appetite and disturbed sleep in workers with the higher urinary mercury levels. Attempts also have been made to study micromercurialism with a psychological test battery.[43,51,66] Two reports note significant relationships between tapping rate and urine mercury.[43,51] Other psychomotor tests did not reveal statistically significant relationships. One test battery[66] included tremor measures, tracking an irregularly moving target with a handle that moved in two dimensions, discriminative reaction time and simple reaction time. No test measure differentiated between exposed chlor-alakali workers (with urinary values of 108 μg/L and whole blood values of 20 μg/L) and workers from a reference group. Details of the scoring and analysis procedures are rather terse and information about the distributions is also lacking, making it difficult to assess the total validity of the study.

3.1.2. CHILDREN

The pervasive incorporation of mercurials into medical treatment left behind a vast legacy of mistreated patients. It also generated many puzzles, largely because the toxic consequences of such treatments were confused with the affliction for which the mercury was prescribed. There is no other way to account for the mystery that acrodynia, or pink disease, presented to pediatricians early in this century. The syndrome is characterized by pink hands and feet, scarlet cheeks and tip of nose, alopecia, excessive salivation and perspiration, rashes, and sometimes desquamation, and marked nervous system disturbances. These include hypesthesias of the extremities accompanied by itching, burning, and severe pain, photophobia, irritability and colicky crying, insomnia, and apathy. Its etiology became the subject of much speculation and debate and was ascribed variously to infectious disease, allergy, endocrine disturbances, gastrointestinal dysfunction, or neuritis.

Warkany and Hubbard[77,78] finally documented the source: mercury. Impressed with the similarities between arsenic poisoning and the salient features of pink disease, they requested chemical determinations of the metal content of urine from children ill with the disease. After finding no relationship with arsenic or the other metals whose analysis they had requested, they were about to discard their hypothesis when a chemist at the Kettering Laboratory in Cincinnati undertook to assay mercury, a metal not on the original list. Some of the samples showed staggering concentrations of mercury. Pursuing this lead, Warkany and Hubbard[78] identified mercury as the source of pink disease. The mercury come from teething powders containing calomel (mercurous chloride), mercury ointments, diapers rinsed in mercuric chloride, and calomel-containing medications prescribed for worm infestations.

Some physicians rejected these data at the time, but subsequent data ultimately confirmed Warkany and Hubbard. In Sheffield, England, for example, the incidence of pink disease declined sharply (Fig. 5) after calomel was eliminated from teething powders.[18] Some of the arguments raised against Warkany and Hubbard originated in the assumption that, if mercury were the source of pink disease, then most of the children treated with teething powders should have become victims. Such critics, of course, failed to realize the enormous variability exhibited by the polymorphic human population in response to any chemical challenge. They also failed to perceive how often mild cases of pink disease must have been misdiagnosed as ordinary childhood illness. Warkany and Hubbard also stressed the critical influence of age, noting that the number of victims less than 4 yr of age was much higher than would be expected from their proportion of the child population.

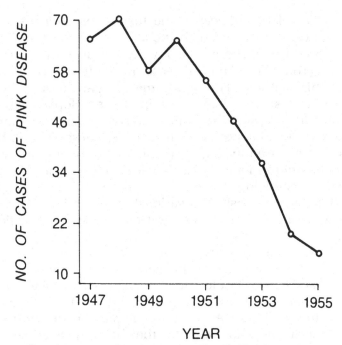

FIG. 5. Decline in cases of Pink Disease in Sheffield, England, after removal of calomel (mercurous chloride) from teething powders.[18]

Pink disease flared despite the lengthy history of mercury poisoning in medicine, and countless descriptions of its manifestations in the literature. Wyllie and Stern[85] remarked, "certainly in the literature no mention of the condition can be found earlier than 1903." Current hazards come from unexpected sources. A survey of infant incubators revealed over one-fourth with mercury vapor levels exceeding the 8-h industrial exposure standard.[76] The source was traced to broken thermometers. The Division of Health and Safety at the University of Rochester (R. Wilson, personal communication) reports that even a single droplet of 0.5 mm in an isolette can give rise to an excessive concentration and described a comprehensive method for eliminating the contamination.

3.1.3. ANIMALS

Inorganic mercury, particularly mercury vapor, has aroused little interest on the part of neurobehavioral toxicology. One conceivable reason is the difficulty posed by the exposure technique, which requires resources both in mercury measurement and in inhalation systems. The first published study in the Western literature[4] was prompted by Soviet claims that their more rigorous

standard for mercury exposure in the workplace derived from
their reliance on central nervous system function rather than tissue damage as the primary criterion of toxicity. Borrowing one of
the techniques fruitfully applied by behavioral pharmacology, the
expermimenters trained pigeons on a multiple reinforcement
schedule. During one component of the schedule, food became
available once every 15 min (FI 15') for a peck on a disk illuminated
by green light. During the second component, food was delivered
for every 60 pecks on the disk illuminated by red light (FR 60). Once
trained, and performing at a stable level, exposure began to 17
mg/m³ for 2 h daily, 5 d/week. As exposure continued, response
rates began to fall (Fig. 6). Latency to onset of this decline differed
substantially among the birds, varying from 3 to 30 weeks. These
behavioral changes preceded by several weeks any overt signs of
intoxication, and gradually faded once exposure ended. A second
study from the same laboratory[7] lowered the exposure concentration to 0.1 mg/m³, the then prevailing occupational standard. The

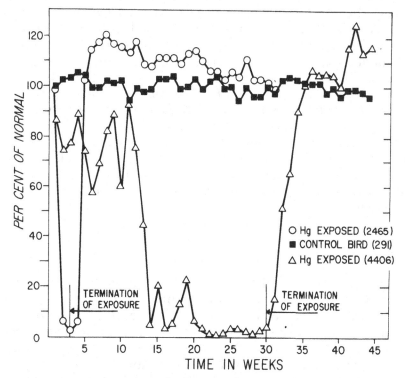

Fig. 6. Response rates on multiple FI/FR reinforcement schedule, expressed as percent of control rates, for two pigeons exposed to mercury
vapor and one to control atmospheres.[4]

authors claim not to have detected any changes in behavior during the 20-week exposure period, but the performance of a control pigeon improved progressively during this time, suggesting that improvement might have served as a better index of adverse effects.

A later rat study[8] showed that exposure to 17 mg/m^3 for 2 h daily impaired avoidance performance in a situation requiring the rat to grip and ascend a pole to avoid an electric shock to the floor of the chamber. Kishi et al.[42] also studied rats trained to avoid or escape electric shocks by climbing a pole. The avoidance group was given a warning signal consisting of a buzzer. The escape group received no warning signal. The experimental groups were exposed for 3 h daily, five times each week, to a concentration of 3 mg/m^3. Significant performance decrements began to appear after about 15–20 weeks of exposure in both experimental groups, and by week 41 all rats showed a 50% decline in responding. The authors noted that impaired performance did not seem to arise out of lethargy or inability to discriminate the warning stimulus because the buzzer provoked defecation, vocalization, and other signs of response. A fine tremor of the head and legs also became apparent during the 18th week of exposure. No distinctive histopathology was observed in brain tissue, and only slight degenerative changes in the tubular epithelium of the kidney. In those rats sacrificed shortly after the onset of depressed performance, kidney mercury levels averaged about 97 µg/g. In the brain, the highest levels appeared in the cerebellum (35 µg/g), with other parts of the central nervous system yielding values of 13–20 µg/g. These values fell by 60% in rats sacrificed 10–11 weeks after the end of exposure, indicating a slow rate of turnover of elemental mercury in the central nervous system, a result consistent with previous reports.[63]

3.2. Methylmercury

Metallic mercury is relatively benign compared to the alkyl compounds of mercury, which are mainly nervous system poisons. Methylmercury is the most potent of these and also the most prevalent because of its presence in fish. It now is known that microorganisms in the bottom sediment of lakes and rivers convert inorganic mercury into the more toxic methyl form. Taken up by plants and plankton, it ascends the food chain to humans. Predatory marine fish, such as tuna and swordfish, also contain substantial levels. Although most of our knowledge of human toxicity has come from disasters, regulatory and research agencies grapple most with the question of risks from fish consumption, and how to set standards for fish burdens that provide adequate protection for the community. Their task is greatly complicated by wide differences between adult and fetal–neonatal susceptibility.

3.2.1. HUMAN ADULT

Although methylmercury had been identified as a potent CNS poison shortly after its synthesis in the middle of the 19th century, the first well-documented human cases were described by Hunter et al.[39] The victims worked in a manufacturing plant producing methylercury fungicide; one was assigned to a chemical laboratory attached to the plant. The sequence of signs and symptoms was consistent. In the laboratory assistant, numbness in fingertips and toes became apparent about 3 months after beginning work. Delicate movements became difficult; it was a struggle to button and unbutton clothes. Coworkers saw a friendly, pleasant boy turn irritable and abusive. Speech labored and sagged. He found it increasingly difficult to detect objects in the periphery of the visual field, and several times narrowly escaped being struck by approaching automobiles. Five weeks later, he began to walk with a staggering, uneven gait and his movements became so clumsy that he could hardly handle a knife and fork. After further deterioration, he began a slow, painful, and incomplete recovery. Twenty-five years later, he still walked with a marked ataxia; a fine, jerky tremor persisted in the upper limbs, head and neck; he still could not distinguish one small object from another by touch; and his vision remained restricted to a small central area of the visual field.[38]

The first intimations of methylmercury as a possible ecological hazard arose in Minamata, the fishing village on the coast of the southernmost Japanese island of Kyushu, whose name is now synonymous with methylmercury poisoning. In the 1950s, the inhabitants of that area were attacked by a central nervous system affliction that came to be known as Minamata Disease. Its source was identified later as methylmercury dumped as effluent into Minamata Bay by a factory that used mercury as a catalyst in acetaldehyde production. Fish and shellfish from the bay took up the methylmercury, and then were eaten by local fishermen and their families. Several hundred cases have been identified up to now. Many died. The rest suffered considerable brain damage. About 10 yr later, a similar episode occurred at Niigata, on the main Japanese island of Honshu, and was traced to a similar source.[72]

The primary signs and symptoms of methylmercury poisoning in adults are listed in Table 3. Central nervous system

Table 3
Indices of Methylmercury Toxicity

Sensory	Motor
Paresthesia	Disturbances of gait
Pain in limbs	Weakness, unsteadiness of legs; falling
Visual disturbances (constriction)	Thick, slurred speech (dysarthria)
Hearing disturbances	Tremor
Astereognosis	*Other*
	Headaches
	Rashes
	"Mental disturbance"

damage predominates. The pattern of overt symptoms from chronic methylmercury poisoning is consistent with the pattern of damage to the brain. Under the microscope, the granule cell population of the cerebellum appears decimated. Masses of neurons are lost from the occipital area of cerebral cortex, particularly in the vicinity of the calcarine fissure, which subserves peripheral vision.[73] Similar destruction, perhaps less pronounced, is discernible in other cortical areas, such as the parietal and temporal lobes.

The most carefully quantified aspect of methylmercury poisoning in adults has been visual function, largely because paresthesia is a subjective manifestation. The episode at Niigata has so far provided most of these data, largely because the Minamata findings prepared Japanese investigators for quickly diagnosing the problem and tracing its development. Figure 7 depicts the visual fields of one of the Niigata victims over a 5-yr period.[40] Progressive constriction developed despite the claims of the patient that he no longer was consuming fish from the contaminated area. This is one question that still lingers. Other evidence suggests partial remission of impairment once exposure stops and blood levels decline (at least over the period of time shown here).

Such constriction is not a peripheral (retinal) effect. Damage to retinal neurons has never been reported. Central nervous system pathology is the source of the visual deficits, especially the lesions invading the calcarine fissure of visual cortex, the area receiving projections from the peripheral portions of the visual field. Peripheral field constriction leads to other problems, such as impaired night vision, because the peripheral field subserves rod function, and rods comprise the light-gathering elements of the visual system.

The ability to track visual targets also is impaired, poisoning victims displaying abnormalities of eye movements as measured

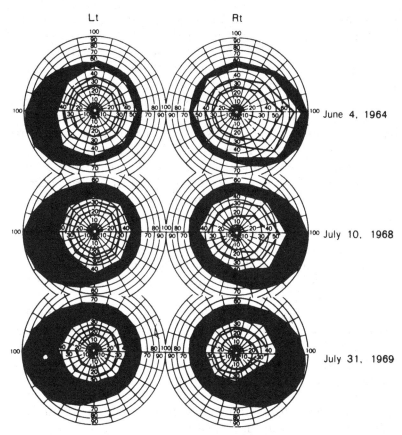

Lt   Rt

FIG. 7.   Progressive constriction of visual fields in methylmercury victim from Niigata, Japan. Black areas depict areas of loss of function.[40]

by the electrooculogram (EOG).[73] In response to targets moving in simple harmonic motion, patients tracked in abrupt steps rather than smoothly. Saccadic pursuit was characterized by overshoot of the target as well as by jerkiness during pursuit. One source of these difficulties probably was cerebellar dysfunction. The cerebellum is a control center for eye movements, and human brains are found to have cerebellar lesions as a consequence of methylmercury intoxication. In addition, areas of the cerebral cortex also subserving this function (frontal eye fields) are damaged by methylmercury.

3.2.2. THE DEVELOPING HUMAN

The Minamata episode, because of its epidemic proportions, also implied a danger not recognized earlier from the occupational exposures.[72] Twenty-three of the 359 children born in villages near the bay between 1955 and 1959 showed symptoms of cerebral palsy, a proportion (6%) much higher than the usual incidence of 0.1–0.6%. The Japanese physicians who interviewed and examined 15 of the mothers claimed that not one, at the time, revealed any clinical symptoms. Reexaminations several years later, however, disclosed neurological abnormalities, consistent with methylmercury poisoning, in 11 of the mothers. Such data suggested that the fetus was considerably more sensitive to methylmercury than the mother, that the placenta was not a barrier to this chemical, and even that the fetus might act as a sink, as it were, for the mother's accumulation of the agent. The pattern of brain damage in infantile Minamata disease is different from the adult pattern (cf. Choi, this volume). Destruction tends to be more uniform. Fetal exposure reduced brain weight in severely poisoned children to one-half or less of normal. Abnormal cells could be observed throughout the brain. The striking specificity of methylmercury for the nervous system is underlined by the absence of detected malformations anywhere else.

An epidemic of poisoning in Iraq yielded the most important body of information about the toxicity of methylmercury to the developing organism. This massive episode occurred in the winter of 1971–1972.[5] Eighty thousand tons of seed grain, mostly wheat, were delivered to Iraqi farmers, but much of it arrived after the planting season. At the government's request, it had been treated with methylmercury fungicide before shipment. Many of the farmers baked the wheat into bread. In the space of about 3 months, deaths reached into the hundreds, probably thousands if unreported deaths are taken into account, and many more victims were afflicted with damage short of death. The scale of the episode, combined with the opportunity to quantify exposure, allowed the first dose–response data to be plotted for humans. Although blood concentrations were measured in many victims, the most important data came from hair analyses. Hair grows at a fairly constant rate (about 1 cm/month), and the ratio of hair methylmercury concentration to blood level is also fairly constant (about 250:1). By chopping hair into 1 cm segments, the history of exposure can be traced month by month.[16]

Twenty-nine mother–infant pairs were selected at first for intensive study. They comprised three groups categorized into low, moderate, and high peak levels.[46] Neurological examinations and interviews with mothers indicated that fetal methylmercury exposure can exert severe, permanent damage leading to mental retardation and other neurological deficits. These findings are espe-

Table 4
Comparison of the EC50s[a] for Adult (Nonpregnant) and Prenatal
Exposures to Methylmercury Compounds Derived from Logit Fit

Adult exposures		Prenatal exposures	
Signs and symptoms	EC50 ± SEM mgHg/kg[a]	Signs and symptoms	EC50 ± SEM mgHg/kg[b]
Paresthesia	2.1 ± 0.41	Motor retardation	0.56 ± 0.19
Ataxia	3.0 ± 0.33	CNS signs	1.01 ± 0.44

[a]The average concentration of methylmercury in the body expresssed as mgHg/kg body weight associated with a 50% frequency of signs and symptoms corrected for background frequency.

[b]The average peak concentration of methylmercury in mothers during pregnancy associated with a 50% frequency of signs and symptoms in their children corrected for background and frequency. The average concentration of methylmercury in the body was calculated by dividing the hair concentration by 175.

cially dramatic because the most common symptom in the mothers during pregnancy was a transient paresthesia; it occurred in 80% of the high exposure group. A larger sample of mother–infant pairs permitted the calculation of more precise dose–response relationships.[17] Such data, based on 76 pairs, are given in Table 4. If the median effective dose for the fetus and adults is compared, the fetus appears several times more sensitive than the adult. But note that this calculation is based on a mild, transient symptom (parasthesia) in the adult, and incapacitating consequences, motor retardation, in the infant.

A further question, arising from exposure during infancy through maternal milk, was examined in 29 children born shortly before the outbreak in Iraq.[3] Women in rural Iraq feed infants during the first year almost exclusively from breast milk, introducing bread during the second year. The mother–infant pairs were studied over a period of 5 yr (1972–1977). Maternal milk concentrations of methylmercury are about 60% of the total mercury in the milk, and the total milk concentrations of mercury are only a small proportion of the blood level. Since infants consume such a large amount of milk, however, and since the developing organism retains methylmercury longer than adults, body burdens can build to high values. The combination of elevated susceptibility and sustained body burdens led to a high incidence of neurological signs in these children. Moreover, the frequency and detectability of these signs increased over the 5-yr period. Hyperreflexia, delayed language development, and delayed motor development were the most characteristic deficits. There also was evidence of mental retardation, but this observation will need to be pursued further as the children mature and can be tested and observed more carefully.

3.2.3. ADULT ANIMALS

Japanese investigators took the lead in neurobehavioral studies of methylmercury intoxication because of the episodes in Japan. Common laboratory species, such as rodents, were the targets of most of this work. Although they may not be suitable for extrapolating the profile of damage or kinetics to humans, some principles and mechanisms have emerged. One is that the unfolding of neurological signs depends not only upon the absolute level of methylmercury in nervous tissue, but upon dosing schedule. Suzuki and Miyama [71] compared the sequence of signs appearing under three dose regimens: 100, 31.5, and 10 μg Hg/g in the diet. The highest concentration produced a cluster of neurological deficits beginning at about day 15, corresponding to a brain level of 30–40 μg/g. With the second highest concentration, the earliest sign, inability to maintain the head in an erect position, appeared several days later and corresponded to a brain level of about 10 μg/g.

Berthoud et al.[11] intubated mice daily with 0, 0.25, 1.0, and 4.0 mg Hg/kg as methylmercury chloride. The animals consumed a liquid diet consisting of evaporated milk supplemented with a vitamin–mineral mixture. They were tested on a battery of motor coordination tasks, and later sacrificed for assays of brain mercury concentrations and histopathology. The first signs to appear were hypophagia and body weight loss, which, even at a daily dose of 1.0 mg Hg/kg, induced such effects at a time (20 d) before gross motor effects were evident. Neurological signs unfolded more gradually in the mice consuming the lower concentrations.

The question of dose rate is central to assessments of hazard because dose rate determines the spectrum of adverse effects associated with intoxication. Investigations in nonhuman primates offer the best measure of that influence because of the functional and structural similarities of all advanced primate brains. Kinetics are also similar. The ratio of brain to blood methylmercury concentration at steady state ranges from 5 to 10 in monkeys, depending upon specific brain region.[30] The value for humans is not notably different.

The impact of dose rate was exemplified in a series of studies at the University of Washington.[33] Methylmercury hydroxide was administered to female rhesus monkeys (Macaca mulatta) at dose rates of 50, 80, 90, 500, and 125 µg/kg/d. Blood levels in the monkeys from the four lowest dose groups reached steady state between 17 and 27 weeks and ranged from 0.9 (50 µg/kg/d) to 1.9 (100 µg/kg/d) µg/mL. The monkeys treated with 125 µg/kg/d never reached steady state. Blood levels continued to rise, suggesting that excretory mechanisms had become saturated and that the kinetics had shifted from first-order to zero-order. None of the monkeys in the two lowest dose groups developed overt signs of neurotoxicity. One animal each from the 90 and 100 µg groups developed such signs. All the monkeys in the highest dose group suffered severe deficits, with the sequence of symptoms compressed into a 3-d period and emerging when concentrations in blood reached 3 ppm.

Both clinical findings and neuropathologic assessment have demonstrated the susceptibility of the human visual system to methylmercury. Nonhuman primates also show prominent effects on vision. Berlin et al.[10] administered oral doses of methylmercury to squirrel monkeys (Saimiri sciurea) trained to discriminate a steady from a flickering light. The flicker frequency remained at 10 Hz, and luminance was varied from low (scotopic) to high (photopic), with criterion defined as the light intensity at which the correct choice was made 70% of the time. During a 6–9 month period, methylmercury was administered at doses designed to achieve a slowly rising blood level. In all six monkeys subjected to this regimen, the first objective sign of toxic disturbance was an increase in the luminance required to detect the flicker. The critical concentration lay near 1.2 µg/g, and it required from about 6 to about 18 months for performance finally to deteriorate substantially. Typically, this final stage developed abruptly, and was accompanied by a sharp rise in blood concentration, as though excretory mechanisms had been overwhelmed.[33] Earlier, however, there were periods, in two of the monkeys, of transient sharp rises in criterion luminance. These periods had been preceded by transient rises in blood levels, although, by the time the performance decrements were manifested, these values had fallen again. A close examination of the figures in this paper suggests that many questions about the kinetics of methylmercury and their relationship to toxic actions on the nervous system remain unanswered.

A more direct means of assessing performance at low luminance was undertaken in this laboratory,[30,31] It was based on the substantial literature in neuropsychology indicating that cortical lesions in primates in those areas subserving vision (area 17) impaired or eliminated the ability of macaque monkeys to discriminate geometric forms. An apparatus was designed to test this ability. Monkeys (*Macaca arctoides*) restrained in a Plexiglas frame faced a panel containing three translucent keys on which could be projected various forms (in this experiment, a square, circle, and triangle). The monkeys were trained to select the key with the square. A few drops of fruit juice were delivered through a spout, for the correct choice, whose position varied randomly. The luminance of the target varied systematically from scotopic levels (requiring 15 min of dark adaptation before the session) to photopic levels. Methylmercury was given orally in biscuits, in doses designed to achieve blood levels in the range that had been associated with clinical visual impairment in Iraq (1.0–2.0 ppm). To reduce the time required to achieve those levels, a series of priming doses was administered at the beginning of the experiment. Weekly maintenance doses were provided to keep blood levels within the targeted range.

One revealing result of this experiment is shown in Fig. 8. At high luminances, accuracy was maintained at 100%, falling to about 80% at dim luminances. After about 10 weeks of treatment, which achieved a blood level of about 3.0 ppm, performance at scotopic levels began to deteriorate. At 20 weeks, dosing was stopped. Performance at high luminances began a sharp decline at this time, but recovered relatively soon. Scotopic performance continued to erode even as blood level fell. This animal, 5 yr since the end of dosing, is in relative good health, but still unable to perform at low levels of luminance, and remains with severely constricted visual fields.

Measurements of visual discrimination performance relying on geometric form, despite their virtue as a functional test, are less useful for dissecting the elements of visual function than those now used by visual scientists. Current methods are based on analyses of visual function that view the system as a Fourier analyzer; that is, a system that decomposes visual scenes into constituent frequency components. These can be temporal (such as flicker) or spatial (light and dark patterning) in character.[47] After developing a method for plotting visual fields in monkeys, Merigan[48] showed how the fields in the monkey whose data are plotted in Fig. 8 were severely constricted. He then demonstrated that this monkey was profoundly impaired in its ability to resolve flicker. Instead of relying on a single frequency,[10] Merigan tested many different frequencies; in addition, he also tested many different values of contrast between the light and dark phases of the flicker (modulation depth). When plotted as a function of different baseline luminances, the resulting thresholds of discrimination yield what are

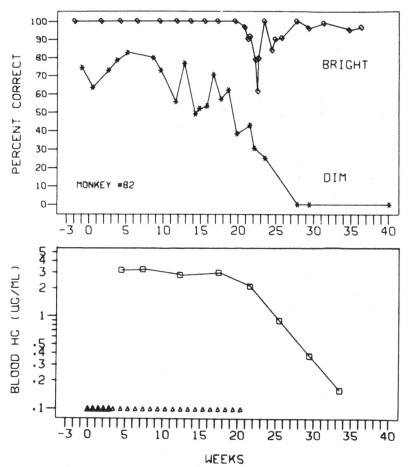

FIG. 8.   Performance on a visual discrimination by a monkey poisoned by methylmercury (top) and corresponding blood levels (bottom). Filled triangles represent priming doses, open triangles represent maintenance doses. Bright and dim refer to luminance of visual stimuli and correspond to photopic and scotopic function, respectively.

termed contrast sensitivity functions. These revealed a striking loss of sensitivity to high frequency flicker at high and middle luminances. At low luminances, the monkey showed profoundly reduced sensitivity at all flicker frequencies.

Spatial contrast sensitivity seemed to detect abnormalities in cynomolgus monkeys (M. fascicularis) dosed from birth with 50 μg/kg/d of methylmercury.[62] Peak blood levels reached 1.2–1.6 ppm, but dropped to a level of 0.6–0.9 ppm after 200 d, when adult diets replaced the infant formula fed up to that time. They were tested between 3 and 4 yr of age. The technique required the monkeys to discriminate between an oscilloscope displaying a vertical sine wave grating (alternating light and dark bands) and one displaying a uniform field of equal average luminance. By varyiung spatial frequency (width of the bands) and contrast (the brightness difference between the light and dark bands), contrast sensitivity functions can be drawn for different average luminances. The experimenters chose two for testing; one at photopic and one at scotopic levels. All five of the treated monkeys showed impaired spatial vision compared to the two controls at scotopic levels; three seemed impaired at photopic levels as well. These findings imply an unusual effect on visual function and should stimulate further exploration.

In contrast to sensory dysfunction, the motor consequences of methylmercury poisoning have not been examined by advanced methods. Overt impairment is only the final stage of a progression. If clinically relevant techniques are to be developed, and correlations with histopathology and neurochemistry investigated, behavioral endpoints need to be chosen that reflect the earlier stages of damage. Motor function can be assessed with sensitive techniques.[32] Figure 9 shows a system constructed for measuring fine motor control in monkeys. The animal is trained to insert a paw into the slot and to apply pressure to a platform mounted on a strain gage whose output is amplified and connected to the analog-to-digital input of a computer. If the monkey maintains a force between specified limits for a specified minimum duration, it is reinforced with a sucrose pellet. Figure 10 traces the development of methylmercury toxicity in one animal. Early toxicity was marked by loss of the ability to apply the appropriate force without overshooting—a relatively subtle impairment. Later, with continued observation and no further dosing, grosser manifestations of toxicity became visible.

Some of the human victims of methylmercury poisoning have complained of relatively subtle, nonspecific problems such as headaches and impaired memory. Such deficits are not addressed by the more narrow focus of sensory and motor function testing. Laties and Evans[44] studied pigeons trained to perform a complex discrimination based on the birds' own behavior. The task required 8 or 9 pecks on a response key followed by one peck on the reinforcement key (both keys were translucent disks). The pigeons were dosed chronically until signs of impairment appeared, then followed as they recovered. Even after all overt signs of toxicity had faded, behavioral deficits remained. They were marked by an in-

Fig. 9. Arrangement for testing motor performance in monkeys (*S. sciurea*). The animal is trained to insert a paw into the slot and to apply pressure on a strain gage. Sucrose pellet is delivered for appropriate responses (e.g., force maintained between 25 and 40 g for 2 s).

creased incidence of response sequences below 8 pecks in length, especially under conditions providing no accessory cues to the completion of the requirement.

3.2.4. DEVELOPING ANIMALS

The experimental counterpart to the human epidemiologic data from Iraq and Minamata provides cogent parallels. The fetal and neonatal sensitivity observed in humans has been confirmed in various species and, in fact, has also been documented once the organisms mature. The earliest experimental data indicating such permanence came from Spyker.[69] She treated pregnant mice with methylmercury at a dose of 8 mg/kg. After birth, the offspring were cross-fostered, then followed for their lifetimes. Animals with obvi-

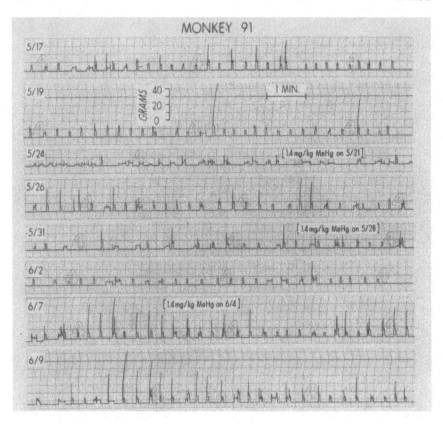

FIG. 10. Strip-chart tracing of motor performance of monkey treated with methylmercury. Pretreatment performance (5/17 and 5/19) shows responses generally reaching within appropriate force range (25–40 g) and remaining there long enough (2 s) to earn reinforcement. Weekly oral methylmercury administration gradually led to disruption of accuracy marked especially by overshoot.

ous defects were discarded. Intact animals, devoid of observable defects, were tested later in various situations. An open-field test revealed a high incidence of backward locomotion in the prenatally treated mice. They also swam in peculiar patterns, another indication of disrupted nervous system development. As they matured, more clinical problems appeared. By middle age (about 18 months) some began to display pustular discharges around the eyes, probably reflecting an immune system defect. Some developed kyphosis and other neuromuscular abnormalities. Some developed obesity. Such aftermaths indicated that, despite overtly normal development, organisms exposed to methylmercury retained lingering nervous system deficits that only rose into visibility wihe reduced neural capacity accompanying aging.

Such deficits may be detectable at much lower doses with appropriate behavioral testing. Musch and his colleagues[12,52] administered methylmercury to pregnant rats on days 6–9 after conception in doses of 0.05 and 2.0 mg/kg/d. Female offspring were trained on an operant task at about 90 d of age. This task, termed differential reinforcement of high rate (DRH), specifies that a prescribed number of responses (lever presses, in this instance) must be emitted within a minimum time to trigger food delivery. For example, one procedure required eight responses within 4 s. For the groups as a whole, response rate and reinforcements fell off in accord with dose of methylmercury. No overt toxicity, such as retarded weight gain, was apparent. The results are not clearly described (for example, no further analyses of IRTs) and indications of variability are not offered. A subsequent study by the same group[13] provided more information. In the latter study, pregnant rats were treated orally with doses of 0.005–0.05 mg/kg methylmercury daily on days 6–9 after conception. Offspring were cross-fostered. Male and female offspring were tested on the DRH schedule when 4 months of age. Dose-related differences in response rates appeared in the procedures requiring four responses in 2 s and eight responses in 4 s. The authors suggested that doses of 0.01 and 0.05 mg/kg/d altered the behavior of both male and female offspring, but the results again were not clearly presented. Alternating 5 min periods during which the schedule was either in effect or not available were combined in the calculations without giving any absolute indication of rates or IRTs. These two studies must be regarded skeptically until proper confirmation is provided. Such poor reporting is especially regrettable because the results, if extrapolated to humans, suggest a threshold dose of methylmercury to rats even lower than that derived for humans on the basis of health disasters.

Another kind of persistent consequence of prenatal treatment was uncovered by drug challenge.[37] Pregnant rats were treated with methylmercury at a dose of 5 mg/kg on gestation days 0, 7, or 14, with no subsequent evidence of toxicity in dams or offspring. Offspring exposed to methylmercury, however, showed reduced sensitivity to the behavioral suppression induced by amphetamine. Such results were confirmed in rats whose mothers had been given single doses of methylmercury (5 or 8 mg/kg) on day 8 or 15 of gestation.[29] Avoidance learning was sharply impaired in the adult offspring of females treated on day 15. Animals whose mothers had been treated on day 8 with 8 mg/kg performed on a spaced responding schedule of reinforcement-like controls. When challenged with 1.0 mg/kg of d-amphetamine, however, they showed far less of a response to the drug (which elevates response rate on this schedule) than control rats.

4. Other Metals

Lead and mercury account for most of the literature on the behavioral toxicity of metals, with lead generating the most intense debates about public policy and biological mechanisms. Yet, as Table 1 demonstrates, other metals also are toxicologically as well as biologically important. They have not, except for scattered reports, received the searching evaluations available through advanced behavioral technology.

This situation seems to be changing as experimenters probe more carefully and with enhanced awareness of important variables such as developmental stage. Cadmium, earlier thought to be almost free of neurotoxicity, is now recognized as damaging to the neonatal nervous system. Intubations of 10 or 100 μg/kg of cadmium chloride during the first 30 d of life increased both locomotor activity and the synthesis and turnover of brain monoamines in rats.[59] Webster and Valois[79] administered cadmium chloride subcutaneously to mice on postnatal days 1, 8, 15, or 22. Damage to the brain was apparent from the treatments on days 1, 8, and 15; the earlier the treatment, the more severe the damage, which was manifested as capillary destruction and hemorrhage. Behavioral deficits were apparent as the animals matured. Similar histopathology appeared in neonatal rats, accompanied by corresponding behavioral abnormalities in a residential maze arranged to measure motor activity.[83] This expanding literature is an exciting development in the behavioral toxicology of metals.

Tin is another metal around which a recent literature has accumulated. Although it had long been recognized that organotin compounds were toxic to humans, new interest was awakened by the recognition that organotins could induce unique types of nervous system lesions. Triethyltin produces edema in the white matter, that is in myelinated tracts, of the brain. Behavioral consequences of such pathology were assessed in rats by measures of forelimb and hindlimb grip strength (measured by a strain gage) and responses to an intense sound and air puff stimulus.[70] Early clinical signs of toxicity were marked by weight loss, general weakness, and insensitivity to painful stimuli. Triethyltin produced dose–related impairment in all of the measures, but recovery took place, once dosing stopped, in a few weeks. Neonates treated with triethyltin on day 5 postnatally, after recovery from the more acute clinical effects, showed enduring behavioral deficits in adulthood and smaller brains.[60]

Two developments led neuroscientists as well to reexamine aluminum as a potential neurotoxicant. One was the discovery that aluminum played a role in the encephalopathy associated with kidney dialysis. The other was the implication[24] that it was associated with senile dementia of the Alzheimer type. The connection with dialysis encephalopathy has been established, although a suitable animal model is lacking. The association with Alzheimer's disease is tenuous, but the issue has stimulated a considerable body of work that both illuminates the biological actions of aluminum and provides further knowledge of degenerative processes in the nervous system. The behavioral data in laboratory animals are ambiguous because, to achieve effects, many experimenters have administered the aluminum intracranially, a mode of administration with obvious pitfalls. A useful summary of the recent literature appears in the volume edited by Liss.[45]

Among the most challenging manifestations of neurotoxicity are those induced by manganese. In mining communities throughout the world, workers have experienced a syndrome, associated with breathing dust from manganese ores, remarkably similar to Parkinson's disease. It can appear within months of the beginning of exposure or require several years for expression, and afflicts only a limited proportion of workers.[53] It also has appeared in other manganese processing settings such as ferro-manganese alloy production, and as inadvertent poisoning through well water contaminated by decaying dry cell batteries.

The manifestations of manganese poisoning, first described in 1837 by Couper, usually are insidious and progressive, and marked by three clusters: psychomotor disturbances, neurological symptoms, and neurological signs.[53] The earliest indications are psychological, and include emotional lability, apathy, and hallucinations. Episodes of unaccountable laughter and a manic type of euphoria may erupt, a phase known to the manganese miners of South America as, locura manganica, or manganese madness. The neurological symptoms are marked by complaints such as weakness, easy fatigability, somnolence, and headache. The clear neurological signs, paralleling Parkinsonism, include gait disorders, speech difficulties, bradykinesia, tremor, ataxia, and dystonias.[19]

The demonstration that these manifestations respond favorably to L-dopa treatment incited the interest of neurochemists and pharmacologists. The literature now contains many demonstrations of manganese localization in various brain areas and of effects on brain catecholamines.[26,28] There is yet to appear, however, a convincing correlation of such neuropharmacologic indices with behavioral variables. Perhaps this problem can be resolved with experiments in primates, because primates possess the pigmented substantia nigra that may be essential for the expression of manganese-induced extrapyramidal disorders.[23]

Of the other metals, only few reports of experimental work are available, although clinical publications abound. There is no doubt that the capacity for neurotoxic damage is more widespread than previously had been believed, but the conditions, such as trace metal balance, supporting such effects need to be defined.

Acknowledgments

The preparation of this chapter was supported in part by grants ES-01247 and ES-01248 from the National Institute of Environmental Health Sciences and in part under Contract No. DE-AC02-76EV03490 with the U.S. Department of Energy at the University of Rochester Department of Radiation Biology and Biophysics and has been assigned Report No. UR-3490-2203.

References

1. Albers, J.W., Potvin, A.R., Tourtellotte, W.W., Peer, R.W. and Stribley, R.F. (1973) Quantification of hand tremor in the clinical neurological examination. *IEEE Trans. Biomed. Eng.* 20,27–37.
2. Angell, N.F. and Weiss (1982) Operant behavior of rats exposed to lead before or after weaning. *Toxicol. Appl. Pharmacol.* 63, 62–71.
3. Amin-Zaki, L., Majeed, M.A., Greenwood, M.R., Elhassani, S.B., Clarkson, T.W. and Doherty, R.A. (1981) Methylmercury poisoning in the Iraqi suckling infant: A longitudinal study over five years. *J. Appl. Toxicol.* 1, 210–214.
4. Armstrong, R.D., Leach, L.J., Belluscio, P.R., Maynard, E.A., Hodge, H.C. and Scott, J.K. (1963) Behavioral changes in the pigeon following inhalation of mercury vapor. *Am. Ind. Hyg. Assoc. J.* 24, 366–375.
5. Bakir, F., Damluji, S.F., Amin-Zaki, L., Murtadha, M., Khalidi, A., Al-Rawi, N.J., Tikriti, S., Dahir, H.I., Clarkson, T.W., Smith, J.C. and Doherty, R.A. (1973) Methylmercury poisoning in Iraq. *Science* 181,230–241.
6. Baloh, R., Sturm, R., Green, B. and Gleser, G. (1975) Neuropsycological effects of chronic asymptomatic increased lead absorption. *Arch. Neurol.* 32, 326–330.

7. Beliles, R.P., Clark, R.S., Belluscio, P.R., Yuile, C.L. and Leach, L.J. (1967) Behavioral effects in pigeons exposed to mercury vapor at a concentration of 0.1 mg/M³. *Am. Ind. Hyg. Assoc. J.* **28**, 482–484.

8. Beliles, R.P., Clark, R.S. and Yuile, C.L. (1968) The effects of exposure to mercury vapor on behavior of rats. *Toxicol. Appl. Pharmacol.* **12**, 15–21.

9. Berlin, M., Fazackerley, J. and Nordberg, G. (1969) The uptake of mercury in the brains of mammals exposed to mercury vapor and to mercuric salts. *Arch. Environ. Health* **18**, 719–729.

10. Berlin, M., Grant, C.A., Heilberg, J., Hellstrom, J. and Schutz, A. (1975) Cerebral cortical pathology, interference with scotopic vision and changes in operant behavior. *Arch. Environ. Health* **30**, 340–348.

11. Berthoud, H.R., Garman, R.H. and Weiss, B. (1976) Food intake, body weight and brain histopathology in mice following chronic methylmercury treatments. *Toxicol. Appl. Pharmacol.* **36**, 19–30.

12. Bornhausen, M., Musch, H.R. and Greim, H. (1980) Operant behavior performance changes in rats after prenatal methylmercury exposure. *Toxicol. Appl. Pharmacol.* **56**, 305–310.

13. Bornshein, R., Pearson, D. and Reiter, L. (1980) Behavioral effects of moderate lead exposure in children and animal models. 2. Animal studies. *CRC Crit. Rev. Toxicol.* **7**, 101–152.

14. Byers, R.K. and Lord, E.E. (1943) Late effects of lead poisoning on mental development. *Amer. J. Dis. Child.* **66**, 471–494.

15. Clarkson, T.W. (1977) Factors involved in heavy metal poisoning. *Fed. Proc.* **36**, 1634–1639.

16. Clarkson, T.W. and Marsh, D.O. (1976) The toxicity of methylmercury in man: Dose-response relationships in adult populations. In: G.F. Nordberg (ed), *Effects and Dose–Response Relationships of Toxic Metals*, Elsevier, Amsterdam, pp. 246–261.

17. Clarkson, T.W., Cox, C., Marsh, D.O., Myers, G.J., Al-Tikriti, S.K., Amin-Zaki, L. and Dabbagh, A.R. (1981) Dose-response relationships for adult and prenatal exposure to methylmercury. In: G.G. Berg and H.D. Maillie (eds), *Measurement of Risks*, Plenum, New York, pp. 111–130.

18. Colver, T. (1956) Pink disease and mercury in Sheffield, 1947–1955. *Brit. Med. J.* **1**, 897–897.

19. Cook, D.G., Fahn, S. and Brait, K.A. (1974) Chronic manganese intoxication. *Arch. Neurol.* **30**, 59–64.

20. Cory-Slechta, D.A. and Thompson, T. (1979) Behavioral toxicity of chronic postweaning lead exposure in the rat. *Toxicol. Appl. Pharmacol.* **47**, 151–159.

21. Cory-Slechta, D.A., Garman, R.H. and Seidman, D.S. (1980) Lead-induced crop dysfunctiuon in the pigeon. *Toxicol. Appl. Pharmacol.* **52**, 462–467.

22. Cory-Slechta, D.A. and Weiss, B. (1982) Delayed behavioral toxicity of lead with increasing exposure concentration: a systematic replication. *Toxicologist* **2**, 81.

23. Cotzias, G.C., Papavasiliou, P.S., Ginos, J., Steck, A. and Duby, S. (1971) Metabolic modification of Parkinson's disease and of chronic manganese poisoning. *Ann. Rev. Med.* **22**, 305–326.

24. Crapper, D.R. and Dalton, A.J. (1973) Alterations in short-term retention, conditioned avoidance response acquisition and motivation following aluminum induced neurofibrillary degeneration. *Physiol. Behav.* **10**, 925–933.

25. Department of Health and Social Security (1980) *Lead and Health.* Her Majesty's Stationary Office, London.

26. Deskin, R., Bursian, S.J. and Edens, F.W. (1981) Neurochemical alterations induced by manganese chloride in neonatal rats. *Neurotoxicology* **2**, 65–73.

27. Dietz, D.D., McMillian, D.E. and Mushak, P. (1979) Effects of chronic lead administration on acquisition and performance of serial position sequences by pigeons. *Toxicol. Appl. Pharmacol.* **47**, 377–384.

28. Donaldson, J., Cloutier, T., Minnich, J.L. and Barbeau, A. (1974) Trace metals and biogenic amines in rat brain. *Adv. Neurol.* **5**, 245–252.

29. Eccles, C.U. and Annau, Z. (1982) Prenatal methylmercury exposure: II. Alterations in learning and psychotropic drug sensitivity in adult offspring. *Neurobehav. Toxicol. Teratol.* **4**, 377–382.

30. Evans, H.L., Garman, R.H. and Weiss, B. (1977) Methylmercury: Exposure duration and regional distribution as determinants of neurotoxicity in nonhuman primates. *Toxicol. Appl. Pharmacol.* **41**, 15–33.

31. Evans, H.L. and Garman, R.H. (1980) Scotopic vision as an index of neurotoxicity. In: W.H. Merigan and B. Weiss (eds), *Neurotoxicity of the Visual System,* Raven Press, New York, pp. 135–147.

32. Falk, J.L. (1970) The behavioral measurement of fine motor control: effects of pharmacological agents. In: T. Thompson, R. Pickens and R.A. Meisch (eds), *Readings in Behavioral Pharmacology.*Appleton-Century-Crofts, New York, pp. 223–236.

33. Finocchio, D.V., Luschei, E.S., Mottet, N.K. and Body, R. (1980) Effects of methylmercury on the visual system of Rhesus Macaque (Macaca Mulatta). I. Pharmacokinetics of chronic methylmercury related to changes in vision and behavior. In: W.H. Merigan and B. Weiss (eds), *Neurotoxicity of the Visual System,* Raven Press, New York, pp. 113–122.

34. Grandjean, P. and Beckmann, J. (1981) Symptoms and signs of lead neurotoxicity. In: S.S. Brown and D.S. Davies (eds), *Organ-directed Toxicity, Chemical Indices and Mechanisms,* Pergamon Press, Oxford, pp. 253–256.

35. Hanninen, H., Hernberg, S., Mantere, P., Vesanto, R. and Jalkanen, M. (1978) Psychological performance of subjects with low exposure to lead. *J. Occupat. Med.* **20**, 683–689.

36. Hanninen, H., Mantere, P., Hernberg, S., Seppalainen, A.M. and Kock, B. (1979) Subjective symptoms in low-level exposure to lead. *Neurotoxicology* **1**, 333–356.

37. Hughes, J.A. and Sparber, S.B. (1978) d-Amphetamine unmasks post-natal consequences of exposure to methylmercury *in utero:* Methods for studying behavioral teratogenesis. *Pharmacol. Biochem. Behav.* **8**, 365–375.

38. Hunter, D. (1969) *The Diseases of Occupations*, Little Brown, Boston, 4th Ed.

39. Hunter, D., Bomford, R.R. and Russell, D.S. (1940) Poisoning by methylmercury compounds. *Quart. J. Med.* **9**, 193–213.

40. Iwata, K. (1980) Neuropathalmologic indices of Minamata Disease in Niigata. In: W.H. Merigan and B. Weiss (eds), *Neurotoxicity of the Visual System*, Raven Press, New York, pp. 165–185.

41. Jason, K.M. and Kellogg, C.K. (1980) Behavioral neurotoxicity of lead. In: R.L. Singhal and J.A. Thomas (eds), *Lead Toxicity*, Urban and Schwarzenberg, Baltimore, pp. 241–271.

42. Kishi, R., Hashimoto, K., Shimizu, S. and Kobayoshi, M. (1978) Behavioral changes and mercury concentrations in tissues of rats exposed to mercury vapor. *Toxicol. Appl. Pharmacol.* **46**, 555–566.

43. Langolf, G.D., Chaffin, D.B., Henderson, R. and Whittle, H.P. (1978) Evaluation of workers exposed to elemental mercury using quantitative tests of tremor and neuromuscular functions. *Am. Ind. Hyg. Assoc. J.* **39**, 976–984.

44. Laties, V.G. and Evans, H.L. (1980A) Methylmercury-induced changes in operant discrimination by the pigeon. *J. Pharmacol. Exper. Therap.* **214**, 620–628.

45. Liss, L. (ed) (1980) *Aluminum Neurotoxicity*, Pathotox, Park Forest South, Illinois.

46. Marsh, D.O., Myers, G.J., Clarkson, T.W., Amin-Zaki, L., Tikriti, S. and Majeed, M.A. (1980) Fetal methylmercury poisoning: clinical and toxicological data on 29 cases. *Ann. Neurol.* **7**, 348–353.

47. Merigan, W.H. (1979) Effects of toxicants on visual systems. *Neurobeh. Toxicol.* **1** (Suppl. 1), 15–22.

48. Merigan, W.H. (1980) Visual fields and flicker thresholds in methylmercury-poisoned monkeys. In: W.H. Merigan and B. Weiss (eds), *Neurotoxicity of the Visual System*, Raven Press, New York, pp. 149–163.

49. Mertz, W. (1981) The essential trace elemenmts. *Science* **213**, 1332–1338.

50. Michaelson, I.A. (1980) An appraisal of rodent studies on the behavioral toxicity of lead: the role of nutritional status. In: R.L. Singhal and J.A. Thomas (eds), *Lead Toxicity*, Urban and Schwarzenberg. Baltimore, pp. 3012–365.

51. Miller, J.M., Chaffin, D.B. and Smith, R.G. (1975) Subclinical psychomotor and neuromuscular changes in workers exposed to inorganic mercury. *Am. Ind. Hyg. Assoc. J.* **36**, 725–732.

52. Musch, H.R., Bornhausen, M., Kriegel, H., and Greim, H. (1978) Methylmercury chloride induces learning deficits in prenatally treated rats. *Arch. Toxicol.* **40**, 103–108.

53. National Research Council (1973) *Manganese*. National Academy of Sciences, Washington, DC.

54. Neal, P.A. et al. (1941) Mercurialism and its control in the felt-hat industry. *US Public Health Bull.* 263.

55. Needleman, H.L., Gunnoe, C., Leviton, A., Reed, R., Peresie, H., Maher, C. and Barrett, P. (1979) Deficits in psychologic and classroom performance of children with elevated dentine lead levels. *N. Eng. J.Med.* 300, 59–65.

56. Patterson, C.C. (1965) Contaminated and natural lead environments of man. *Arch. Environ. Health* 11, 344–360.

57. Pentschew, A. and Garro, F. (1966) Lead encephalo-myelopathy of the suckling rat and its implications on the porphyrinopathic nervous diseases with special reference to the permeability disorders of the nervous system's capillaries. *Acta Neuropath.* 6, 266–278.

58. Perino, J. and Ernhart, C.B. (1974) The relation of subclinical lead level to cognitive and sensorimotor impairment in black preschoolers. *J. Learning Disab.* 7, 26–30.

59. Rastogi, R.B., Merali, Z. and Singhal, R.L. (1977) Cadmium alters behavior and biosynthetic capacity for catecholamines and serotonin in neonatal rat brain. *J. Neurochem.* 28, 789–794.

60. Reiter, L.W., Heavner, G.B., Dean, K.R. and Ruppert, P.H. (1981) Developmental and behavioral effects of early postnatal exposure to triethyltin in rats. *Neurobehav. Toxicol. Teratol.* 3, 285–293.

61. Rice, D.C., Gilbert, S.G. and Willis, R.F. (1979) Neonatal low-level lead exposure in monkeys: Locomotor activity, schedule-controlled behavior, and the effects of amphetamine. *Toxicol. Appl. Pharmacol.* 51, 503–513.

62. Rice, D.C. and Gilbert, S.G. (1982) Early chronic low-level methylmercury poisoning in monkeys impairs spatial vision. *Science* 216, 759–761.

63. Rothstein, A. and Hayes, A. (1964) The turnover of mercury in rats exposed repeatedly to inhalation of vapor. *Health Phys.* 10, 1099–1113.

64. Rutter, M. (1980) Raised lead levels and impaired cognitive/behavioral functioning:A review of the evidence *Develop. Med. Child Neurol.* 22, 1–26.

65. Sauerhoff, M.W. and Michaelson, I.A. (1973) Hyperactivity and brain catecholamines in lead-exposed developing rats. *Science* 182, 1022–1024.

66. Schuckmann, F. (1979) Study of preclinical changes in workers exposed to inorganic mercury in chloralkali plants. *Int. Arch. Occup. Environ. Hlth.* 44, 193–200.

67. Silbergeld, E.K. and Goldberg, A.M. (1975) Pharmacological and neurochemical investigations of lead-induced hyperactivity. *Neuropharm.* 14, 431–444.

68. Smith, R.G., Vorwald, A.J., Patil, L.S. and Mooney, T.F. (1970) Effects of exposure to mercury in the manufacture of chlorine. *Am. Ind. Hyg. Assoc. J.* 31, 687.

69. Spyker, J.M. (1975) Behavioral teratology and toxicology. In: B. Weiss and V.G. Laties (eds), *Behavioral Toxicology*, Plenum, New York. pp. 311–344.

70. Squibb, R.E., Carmichael, N.G. and Tilson, H.A. (1980) Behavioral and neuromorphological effects of triethyltin bromide in adult rats. *Toxicol. Appl. Pharmacol.* **55**, 188–197.

71. Suzuki, T. and Miyama, T. (1971) Neurological symptoms and mercury concentration in the brain of mice fed with methylmercury salt. *Ind. Health* **9**, 51–58,

72. Tsubaki, F. and Irukayama, K. (eds) (1977) *Minamata Disease*, Elsevier, New York.

73. Tsutsui, J. (1980) Clinical and pathological studies of eye movement disorders in Minamata disease. In: W.H. Merigan and B. Weiss (eds), *Neurotoxicity of the Visual System*, Raven Press, New York, pp. 187–202.

74. Valciukas, J.A., Lilis, R., Eisinger, J., Blumberg, W.E., Fischbein, A. and Selikoff, I.J. (1978) Behavioral indicators of lead neurotoxicity:results of a clinical survey. *Int. Arch. Occup. Environ. Health* **41**, 217–236.

75. Van Gelder, G.A., Carson, T., Smith, R.M. and Buck, W.B. (1973) Behavioral toxicologic assessment of the neurologic effect of lead in sheep. *Clin. Toxicol.* **6**, 405–418.

76. Waffarn, F. and Hodgman, J.E. (1979) Mercury vapor contamination of infant incubators: A potential hazard. *Pediatrics* **64**, 640–642.

77. Warkany, J. and Hubbard, D.E. (1948) Mercury in the urine of children with acrodynia. *Lancet* **1**, 829–830.

78. Warkany, J. and Hubbard, D.M. (1951) Adverse mercurial reactions in the form of acrodynia and related conditions. *Am. J. Dis. Child.* **81**, 335–373.

79. Webster, W.S. and Valois, A.A.(1981) The toxic effects of cadmium on the neonatal mouse CNS. *J. Neuropath. Exper. Neurol.* **40**, 247–257.

80. Weiss, B. (1970) The fine structure of operant behavior during transition states. In: W.N. Schoenfeld (ed), *The Theory of Reinforcement Schedules*, Appleton-Century-Crofts, New York, pp. 277–311.

81. Weiss, B. and Gott, C.T. (1972) A microanalysis of drug effects on fixed-ratio performance in pigeons. *J. Pharmacol. Exp. Ther.* **180**, 189–202.

82. Weiss, B. (1978) The behavioral toxicology of metals. *Fed. Proc.* **37**, 22–27.

83. Wong, K.L. and Klaassen, C.D. (1982) Neurotoxic effects of cadmium in young rats. *Toxicol. Appl. Pharmacol.* **63**, 330–337.

84. Wood, R.W., Weiss, A.B. and Weiss, B. (1973) Hand tremor induced by industrial exposure to inorganic mercury. *Arch. Environ. Health* **26**, 249–252.

85. Wyllie, W.G. and Stern, R.O. (1931) Pink disease: Its morbid anatomy with a note on treatment. *Arch. Dis. Child* **6**, 137–156.

86. Yule, W., Landsdown, R., Millar, I.B. and Urbanowicz, M.A. (1981) The relationship between blood lead concentrations, intelligence and

attainment in a school population: A pilot study. *Develop. Med. Child. Neurol.* **23**, 567–576.

87. Zenick, H., Rodriquez, W., Ward, J. and Elkington, B. (1979) Deficits in fixed-interval performance following prenatal and postnatal lead exposure. *Develop. Psychobiol.* **12**, 509–514.

Chapter 2

Cadmium Toxicity and Neural Cell Damage

Björn Arvidson

1. Introduction

Cadmium was discovered in 1817 by the German investigator Friedrich Stromeier.[124] The name cadmium was taken from the ancient Greek word for zinc oxide.[8] Toxic effects of cadmium in mammals were reported as early as 1858 by Sovet[109] and 1867 by Marmé.[76,91] An extensive literature has now accumulated on the adverse effects of cadmium on various tissues in man and animals (see reviews by Friberg et al.,[33,34] Fassett,[23,24], Fleischer et al.,[27] Samarawickrama[98]).

In the middle of the 1960's, Gabbiani[35] reported that a single subcutaneous injection of cadmium chloride produced hemorrhagic lesions and nerve cell necrosis in sensory ganglia in rats. Later investigators have also demonstrated the toxic effects of cadmium in other regions of the nervous system, including the brain, autonomic ganglia, peripheral nerves, and the neuromuscular junction. This information is based almost entirely on experimental animal studies, whereas very little is known about possible toxic effects of cadmium on the human nervous system.

In this chapter, some attention will be paid initially to the chemistry of cadmium and to the sources of cadmium pollution.

Björn Arvidson: Neuropathological Laboratory, Department of Pathology, University of Uppsala, Uppsala, Sweden

Thereafter, a survey will be made of reports concerning neural cell damage following exposure to cadmium. Short reviews on the neurotoxicology of cadmium have recently been published by Tischner[112] and Arvidson.[6]

2. Physical and Chemical Properties

Cadmium is a transition metal in group IIB of the periodic table of elements. The metal is bluish-white to silver-white. At room temperature, it has a hexagonal close-packed crystal structure. Eight stable isotopes are known to be present in nature.[8,27] The atomic weight of cadmium is 112.4 and the atomic number 48. The density at 25°C is 8.6 g/cm[3], the melting point 321°C and the boiling point 765°C. The most common oxidation state is +2.[108] The most important compounds are cadmium acetate, cadmium sulfide, cadmium sulfoselenide, cadmium stearate, cadmium oxide, cadmium carbonate, cadmium sulfate, and cadmium chloride. The acetate, chloride, and sulfate are soluble in water, whereas the oxide and sulfide are almost insoluble.[34]

When cadmium is oxidized by air, a thin greyish-white film is formed that protects the metal from further attack. In air, the powdered metal burns with a red flame. Cadmium is not affected by alkalis, whereas dilute mineral acids dissolve the metal, with evolution of hydrogen.[8]

3. Occurrence, Industrial Uses, and Routes of Exposure

In nature cadmium is found together with zinc and is obtained as a byproduct in extraction processes for zinc and other metals.[33] The cadmium zinc ratio is between 0.01 and 0.001 in most minerals and soils. The average content of cadmium in the earth's crust has been estimated at 0.1–0.2 ppm. In 1977, the total production in the world was about 18,000 tons.[8]

Cadmium is used industrially as a protective coating for iron, steel, and copper by electroplating. Cadmium sulfide and sulfoselenide are used as pigments in plastics, enamels, and paint. It may also serve as an alloy with copper for coating telephone cables, trolley wires, and welding electrodes. The stearate is used as a stabilizer in plastics, and cadmium electrodes are found in alkaline accumulators. Cadmium is also used for the production of control rods in nuclear reactors, in semiconductor and photovoltaic devices, and as a fungicide.[8,24,33,34]

Human beings are exposed to cadmium in food, water, and air. Of these routes of exposure, food is considered to be the major one. In contaminated areas, cadmium is accumulated in certain kinds of food, especially liver and kidney from animals and shellfish.[56] High concentrations may also be found in oysters[90] and crabs.[94] Rice and wheat in exposed areas may contain high levels of cadmium[77]. The daily intake of cadmium from food in uncontaminated areas has been estimated at 25–60 μg for a 70 kg person.[33]

In natural water, the cadmium concentration is usually less than 1 μg/L in nonpolluted areas.[34] Contamination of the water may occur from metal or plastic pipes or from industrial discharges.[33,105]

In ambient air, cadmium occurs in particulate form, probably mainly as cadmium oxide.[34] Concentrations of cadmium in air normally range from 0.001 to 0.05 μg/m^3. Higher values have been recorded in cities with metallurgical and smelting industries. Tobacco smoking increases the body content of cadmium since each cigarette contains 1–2 μg of cadmium[71,111] It has been estimated that smoking of two packs of cigarets a day will cause inhalation of 4–6 μg of cadmium.[27]

4. Metabolism

The body content of cadmium increases with age. In the newborn, the total body content is less that 1μg, increasing to about 15–20 mg in an adult human in the UK and Sweden.[33] Cadmium enters the body mainly via the respiratory and gastrointestinal tracts. According to experimental data, after inhalation about 10–40% is absorbed.[33,98] The absorption depends on the particle size and solubility, with higher values for small and highly soluble particles.[34]

After ingestion, an absorption of about 2% has been found in animals,[21,1,0] whereas an average absorption of about 6% was noted in five human beings given single doses of radioactive cadmium.[92] Calcium, iron, or protein deficiency increases the retention rate.[26,68,110]

Cadmium that has been absorbed from the lungs or the intestines is transported in the blood to other organs. After chronic exposure in human beings, about 50% of the cadmium is deposited in the liver and kidneys[27,33,78,121,122] and high concentrations are also found in the pancreas and the salivary glands.[33,34] At a cadmium concentration in the renal cortex of about 200 μg/g wet weight, proteinuria occurs as a result of tubular dysfunction.[33] When the kidneys have been damaged by cadmium, the urinary excretion of the metal increases and at the same time the kidney level of cadmium decreases considerably.[33,78]

After intravenous injection in experimental animals, cadmium in the blood is initially found mainly in the plasma, but later a shift takes place to red blood cells, where the cadmium binds to proteins such as metallothionein and hemoglobin.[73,78,80,86] Metallothionein is of great importance for the transport, excretion, and toxicity of cadmium. It was initially isolated from equine renal cortex[60,75] and has a molecular weight of about 6000–7000. It can bind cadmium, zinc, and other metals owing to the presence of a large number of sulfhydryl groups and has been isolated from several tissues in man and animals.

The excretion of cadmium via the urine and feces is very low in normal human beings and the biological half-life is probably between 10 and 30 yr for the total body.[33]

5. General Toxicology

5.1. Acute Intoxication

Ingestion of cadmium causes severe but temporary gastroenteritis, with nausea, vomiting, abdominal pains, diarrhea, headache, and salivation.[16] Acute intoxication in rats through inhalation of finely dispersed cadmium oxide led to acute pulmonary edema within 24 h of exposure and a proliferative interstitial pneumonitis from the 3rd to the 10th d after exposure.[84] Damage to the kidneys with transient anuria has been reported in humans after intake of cadmium iodide,[125] and in animals, tubular lesions have been described after cadmium injection.[29,62] Damage to the liver after intravenous administration of cadmium included cell necrosis, congestion, and vacuolation of Kupffer cells.[54]

Subcutaneous injections of cadmium salts in animals result in capillary stasis, edema, and hemorrhages in the testis[82,83] These events are followed by regressive changes of the seminiferous epithelium 4–6 h after injection and total necrosis within 24–48 h. The vascular lesions are morphologically similar to those in peripheral sensory ganglia.[40]

5.2. Chronic Intoxication

Inhalation of cadmium oxide fumes, cadmium oxide dust, and cadmium pigment dust for prolonged periods of time can produce emphysematous changes in the lungs.[30,31,118] The exposure time has generally been several years. In animals exposed to cadmium dust the lungs showed interstitial pneumonia, sclerosis, and emphysema.[118]

Disturbed renal function, with proteinuria, glucosuria, and aminoaciduria, has been observed in workers exposed to cadmium.[31] The proteinuria in chronic cadmium poisoning is of the tubular type and is caused by decreased reabsorption of proteins.[20,50] As mentioned previously, when the kidneys have been damaged by cadmium, the excretion of the metal increases and the kidney concentration of cadmium will diminish considerably.[33]

A relation between cadmium exposure and hypertension has been suggested[14,102,103] but the data are still inconclusive.[33,69]

Long-term exposure to cadmium can cause bone changes, including osteomalacia and resulting in pseudofractures. Cadmium does not exert its effect directly on bone tissue, but disturbs the regulation of the calcium and phosphorus balance via renal tubular dysfunction, which in turn leads to skeletal damage. Itai-itai disease, a bone disease reported from Japan, is thought to be caused by chronic cadmium poisoning combined with certain dietary deficiencies, i.e., lack of calcium and vitamin D.[33,34,47,48]

Other effects reported in humans after chronic exposure to cadmium are anemia, liver disturbances, and yellowing of the dental enamel.[33,34,69] Teratogenic effects have been observed in long-term studies on mice, rats, and hamsters.[25,104]

5.3. Carcinogenic Effects

Subcutaneous injections of cadmium salts in rats induce sarcoma at the site of injection and may also give rise to interstitial tumors in the testis.[43-46] In human beings, cadmium has been suggested as an etiologic factor in the development of prostatic carcinoma[63,64,88,89] and cancer of the respiratory tract.[70]

6. Neurotoxicology

6.1. Human Cadmium Neurotoxicology

Very few investigators have reported toxic effects of cadmium on the human nervous system. It has long been known, however, that inhalation of cadmium dust may cause anosmia.[1,9,30,31] The anosmia is reversible if exposure to cadmium dust is discontinued.[9] The mechanism underlying this effect is still not entirely clear. Chronic irritative changes [1,31] and ulcerous lesions[9] are often found in the olfactory mucosa of cadmium-exposed workers and a plausible explanation for the anosmia is that the peripheral olfactory receptors are damaged in this process. The severity of mucosal damage is of no value, however, in assessing the degree of hyposmia.[1] Baader[9] reports post mortem findings in a worker exposed to cadmium oxide dust for 16 yr. The nasal mucosa were atrophic and, interestingly, the olfactory bulbs were stained intensively yellow. The latter observation suggests that cadmium may exert effects not only on the peripheral olfactory receptors, but also on the central olfactory connections.

Wood[128] has suggested as an alternative explanation that cadmium-induced anosmia may result from the displacement of zinc, since cadmium is an antagonist to zinc and zinc deficiency can result in disorders of smell and taste.

The cadmium content of the central nervous system (CNS) has been measured in workers dying after prolonged exposure. These studies have shown very low concentrations of cadmium in the brain and spinal cord.[31,32,107] The findings from investigations of various regions of the brain by light microscopy have been considered normal[31,67] The results of these studies thus indicate that normally the blood-brain barrier protects against penetration of cadmium into the brain parenchyma.

Symptoms from the CNS were recorded in one study of 160 workers in an accumulator factory. The symptoms were partly unspecific and consisted of headache, vertigo, and sleep disturbances. Physical examination revealed exaggerated knee-joint reflexes, tremor, dermographia, and sweating. In tests of sensory, dermal, and motoric chronaxia, the results were abnormal in cadmium-exposed workers with subjective disturbances.[119]

Baader[9] made some interesting findings in a postmortem study of a worker exposed to cadmium oxide dust for 16 yr. The intestines and stomach were segmentally dilated and microscopic examination showed regressive changes and destruction of ganglion cells in the plexus myentericus of the gut. Similar changes were observed in ganglion cells of the stomach, lower esophagus, and pars membranacea of the bronchial and tracheal walls. A systematic study of autonomic ganglia (sympathetic thoracic, 3rd–7th cervical and stellate ganglia and also the nuclei of the vagus nerve and the region of the third ventricle) gave negative results. Sensory ganglia were not included in this study.

6.2. Laboratory Animals

6.2.1. PERIPHERAL NERVOUS SYSTEM

6.2.1.1. Toxic Effects of Cadmium on Peripheral Sensory and Autonomic Ganglia.

In 1966, Gabbiani[35] accidentally discovered that subcutaneous administration of cadmium chloride to rats in doses of 2.65–29 mg/kg body weight (bw) produced hemorrhagic lesions in the dorsal root and Gasserian ganglia. Within the ganglia, the hemorrhages were distributed around damaged nerve cells. The latter exhibited nuclear pycnosis or karyorrhexis, with lysis of the cytoplasm (Fig. 1). Cadmium-induced hemorrhages in sensory ganglia were also seen in other species, including Guinea-pigs, golden hamsters, and mice. Pretreatment of the animals with zinc, glutathione, or cobalt chloride prevented these lesions.[38,39]

When cadmium injections were repeated, the animals developed tolerance against its neurotoxicologic effects.[38] In rats hemorrhagic lesions in ganglia were not seen before the age of 20 d, whereas in the rabbit the lesions were only observed in young animals.[37]

Electron microscopy of sensory ganglia after acute cadmium intoxication has shown that the endothelial cells are damaged at

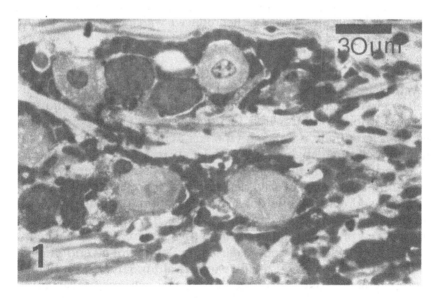

FIG. 1. Section from the trigeminal ganglion of a mouse injected iv with cadmium in a dose of 5 mg/kg bw and killed after 24 h. Dark areas correspond to hemorrhages around nerve cells. Several nerve cells are necrotic.

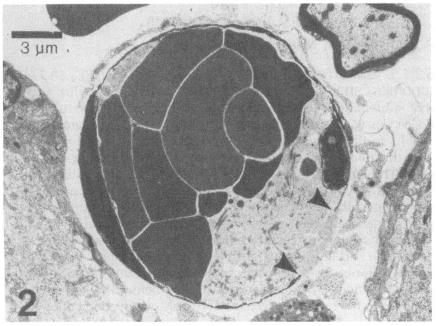

FIG. 2. Electron micrograph of a blood vessel in the trigeminal ganglion of a mouse injected iv with cadmium in a dose of 5 mg/kg bw and killed after 24 h. The endothelial cell cytoplasm shows degenerative changes, with increased density and discontinuities (between arrow heads). The nucleus is small and pycnotic. Reproduced with permission of *Acta Neuropathologica*.

an early stage. Gabbiani et al.[40] found that 15–30 min after an intravenous injection of cadmium chloride in rats, the intercellular clefts were dilated and the endothelial cell membrane was injured. These authors reported that in rat Gasserian ganglia, arterioles were preferentially damaged, whereas Schlaepfer[100] claimed that in rat dorsal root ganglia the main effect was exerted on capillaries and venules. During the first hours after exposure, the endothelial cells underwent progressive degenerative changes, characterized by increased cytoplasmic density, vacuolation, nuclear pycnosis, and fragmentation of the endothelial lining (Fig. 2). Red blood cells and plasma proteins passed through the gaps in endothelial cells, leading to interstitial hemorrhages and edema. Not all endothelial cells in the ganglia were sensitive to cadmium. Thus, in rat dorsal root ganglia, at least 25% of the endothelial cells were resistant.[100] The selective survival of cadmium resistant endothelial cells may explain the tolerance against a second dose of cadmium.[100]

Arvidson[5] studied the distribution of hemorrhagic lesions in ganglia of mice following a single subcutaneous injection of cadmium chloride. Hemorrhages and nerve cell necrosis were constant findings in sensory ganglia. In the superior cervical ganglion, focal hemorrhages occurred in about 50% of the treated animals, whereas no vascular lesions or nerve cell damage were observed in the celiac ganglion. These pronounced variations in the extent of tissue damage between various types of ganglia have not yet been explained.

Very little is known about the pathogenesis of cadmium-induced vascular injury in ganglia. The lesions are similar to those that occur in the testis[40] and the mechanism of the endothelial cell damage is probably the same. In this context, it is interesting that similarities have been found betwen the ultrastructure of testicular and ganglionic endothelial cells. In both testes and ganglia, the endothelial cells are characterized by a large number of microvilli projecting from the luminal surface.[36]

Functional differences, such as variations in the occurrence of certain enzymes, may exist between different vascular beds.[97,116] In brain capillary endothelial cells, the distribution of alkaline phosphatase and $Na^+,K^+ATPase$ differs between the luminal and antiluminal cytoplasmic membranes.[11] Cadmium is known to inhibit several SH-containing enzymes[117] and it might be speculated that the content and subcellular distribution of enzymes in ganglionic endothelial cells are of importance for the preferential occurrence of cadmium-induced vascular injury in ganglia. A similar theory has been proposed to explain the vascular damage caused by cadmium in the testis. Thus, Aoki and Hoffer[2] suggested that a cadmium-sensitive testicular isoenzyme of carbonic anhydrase may in some way be associated with the capillary endothelial cells.

The reason for the variation in sensitivity to cadmium between individual endothelial cells in sensory ganglia is not completely understood. Schlaepfer[100] has suggested a number of possible explanations, such as differences in metabolic activity, in cell membrane charge or porosity, and in intrinsic chelation mechanisms between endothelial cells.

Blood vessels in peripheral sensory and autonomic ganglia are fenestrated and highly permeable to proteins.[4,57,58,81] After intravenous injection, cadmium binds to various plasma proteins and only very small amounts cross the blood-brain barrier, whereas there is pronounced accumulation in the ganglia.[7] This concentration of cadmium in ganglia, which can be explained by the unique permeability properties of ganglionic vessels, is presumably of importance for the development of hemorrhagic lesions in these parts of the nervous system.

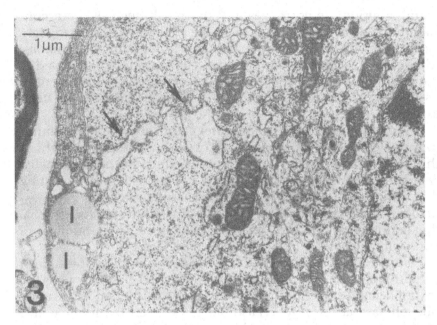

FIG. 3. Nerve cell from the trigeminal ganglion of a mouse injected iv with cadmium in a dose of 5 mg/kg bw and killed after 24 h. The cisternae of the rough endoplasmic reticulum are dilated (arrows) and mitochondria display increased matrical density. Ribosomes are lying freely in the cytoplasm. Lipid vacuoles (l) are present in a satellite cell. Reproduced with permission of *Acta Neuropathologica*.

The cadmium-induced hemorrhagic lesions in ganglia are accompanied by damage to nerve cells. In a study of ganglionic lesions in mice,[5] slightly damaged neurons displayed dilatation of the endoplasmic reticulum, condensation of the mitochondrial matrix, and accumulation of lipid in the cytoplasm (Fig. 3). In more severely damaged nerve cells, the cisternae of the rough endoplasmic reticulum were fragmented and the organelles were assembled in the perinuclear region. Nerve cell necrosis was a common finding (Fig. 4). Myelinated axons in trigeminal and dorsal root ganglia were pathologically altered, with proliferation and dilatation of the smooth endoplasmic reticulum and accumulation of lamellated electron-dense bodies in the axoplasm (Fig. 5).

Most authors agree that the damage to nerve cells in ganglia after acute cadmium intoxication is secondary to the vascular lesions.[39,40,100] In vivo, it is almost impossible to discriminate the direct toxic actions of cadmium on nerve cells from the effects of vascular damage. In tissue cultures, however, the vascular effects

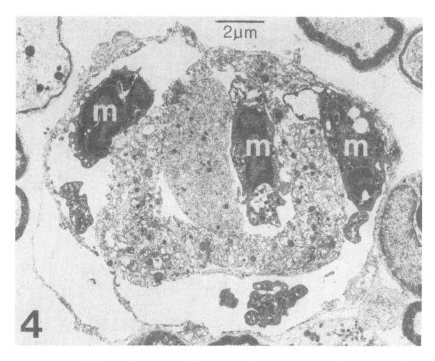

Fig. 4. Electon micrograph from the trigeminal ganglion of a mouse killed 72 h after iv injection of cadmium in a dose of 5 mg/kg bw. Macrophages (m) are phagocytosing the remnants of a necrotic nerve cell. Reproduced with permission of *Acta Neuropathologica*.

are eliminated, and such studies clearly demonstrate that cadmium also exerts a direct toxic effect on nerve cells. Thus, in tissue cultures from rat dorsal root ganglia, neurons accumulated glycogen, lipid droplets, and masses of neurofilaments in the cytoplasm. Unmyelinated axons were focally distended and in these regions degenerating mitochondria and myelin figures were observed.[113] On exposure of chick dorsal root ganglia to cadmium in another tissue culture study the result was inhibition of cell growth. Nerve fibers were more sensitive to cadmium than neuroglia and neuronal cell bodies. The supporting neuroglia was pathologically altered, with vacuolation, granulation, atrophy, and lytic changes.[106]

6.2.1.2. Peripheral Nerves. Rats exposed to 10–40 ppm of cadmium chloride in their drinking water for 18–31 months developed peripheral neuropathy, with weakness in the hindlimbs and muscle atrophy. The main finding was myelin degeneration, phag-

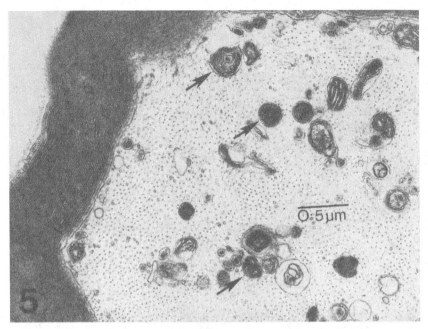

FIG. 5. Myelinated axon from the trigeminal ganglion of a mouse killed 48 h after iv injection of cadmium in a dose of 5 mg/kg bw. Several lamellated electron-dense bodies (arrows) are seen in the axoplasm.

ocytosis of myelin, and remyelination. Another prominent finding was accumulation of glycogen particles in glycogenosomes of the axoplasm.[99] The mechanism of glycogenosome formation in axons has since been studied in detail by the same authors.[49] Cadmium inhibited the activity of neutral α-glycosidase in rat sciatic nerve, and this inhibition is likely to cause enlargement of glycogen particles, which are subsequently engulfed in the lysosome system.[49] In acute cadmium intoxication in mice, vascular lesions similar to those in ganglia may also occur in peripheral nerves. This damage is characterized by endoneurial edema with interstitial hemorrhages and endothelial cell degeneration.[5]

In a recent study, solutions of lead acetate, cadmium chloride, and mercuric chloride were injected intraocularly to study the effect of these salts on the sympathetic adrenergic ground plexus of the iris, using fluorescence histochemistry. Injection of lead caused a significant hyperinnervation compared with controls, whereas mercury chloride resulted in marked degeneration of adrenergic fibers. Cadmium chloride had no effect on the adrenergic nerve fiber density, but caused local irritative effects with bleeding, swelling of the iris, and a macrophage reaction.[12]

Perineurial cells have some features in common with endothelial cells, namely the frequency of pinocytotic vesicles, the presence of a basement membrane and similar junctions between adjacent cells. Acute cadmium intoxication in mice did not, however, produce any degeneration of perineurial cells in nerves or peripheral ganglia and the perineurial diffusion barrier was intact also in cadmium-treated animals.[5]

6.2.1.3. Neuromuscular Transmission.

Cadmium ions may inhibit neuromuscular transmission, a feature shared with several other divalent cations. This effect is presumably caused by inhibition of calcium function at presynaptic nerve terminals, with a consequent reduction of the release of acetylcholine from the nerve endings.[17,28,52,115] Toda[114] has suggested that Cd^{2+} may interfere with the influx of Ca^{2+} by binding to sulfhydryl groups in membranes. From a comparison of the effects of cadmium and lead on adrenergic neuromuscular transmission, Cooper and Steinberg[17] concluded that cadmium at any dose was a more potent blocking agent than lead. Cooper et al.[18] propose that cadmium and lead may exert their effect by different mechanisms. They suggest that lead inhibits the passage of Ca^{2+} through the presynaptic membrane, whereas cadmium interferes with binding of Ca^{2+} to some intracellular site.

6.2.2. CENTRAL NERVOUS SYSTEM

6.2.2.1. Distribution of Cadmium in the Central Nervous System After Parenteral or Oral Administration.

Walsh and Burch[120] studied the distribution of ^{115m}Cd in normal dogs after intravenous injection and found no uptake in the brain, phrenic nerve, or brachial plexus up to 48 h after the injection. Autoradiographic studies with intravenously administered ^{109}Cd revealed only traces of cadmium in the brain parenchyma, whereas pronounced accumulation was seen in the pia mater, choroid plexus, and hypophysis.[10,79] Within the pituitary, the pars intermedia contained less cadmium than the rest of the gland.[10] Arvidson and Tjälve[7] found accumulation of ^{109}Cd also in the pineal gland of rats 1 week after intravenous injection (Figs. 6, 7).

Buhler et al.[13] investigated the tissue distribution of $^{109}CdCl_2$ supplied with food or drinking water for 1–2 weeks in rats. The percentage of the body content that accumulated in brain varied from 0.0017 to 0.071. In another study, in which ^{115}Cd was supplied with the drinking water, the accumulation of cadmium in the brain was surprisingly high. Thus, after addition of ^{115}Cd to the drinking water for 8 d in rats, the uptake in the brain was

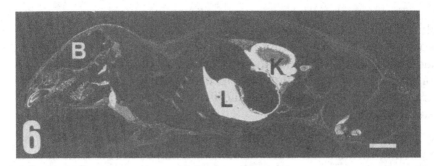

FIG. 6. Autoradiogram of a sagittal whole-body section of a rat killed 1 week after iv injection of 50 μCi of [109]CdCl$_2$. Light areas correspond to high uptake of cadmium. The metal has accumulated in the liver (L) and kidney (K), whereas no uptake is seen in the brain parenchyma (B).

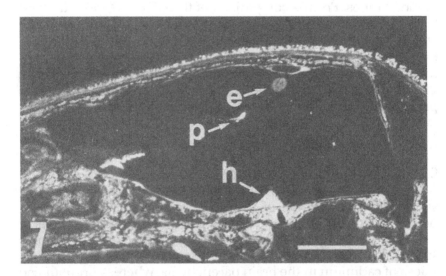

FIG. 7. Detail of autoradiogram from a rat killed 1 week after iv injection of 50 μCi of [109]CdCl$_2$, showing a sagittal section of the brain. There is no detectable uptake in the brain parenchyma, but cadmium has accumulated in the hypophysis (h), choroid plexus (p) and epiphysis (e). Uptake of cadmium is also seen in the meninges.

found to be about 20% of that in the liver. For obscure reasons, the accumulation in the striatum was significantly higher than in the cerebellum or cerebral cortex.[53]

Cadmium accumulated in the brains of fetuses when pregnant rats were given a subcutaneous injection of [109]Cd. The extent of this accumulation was not specified, although less cadmium was found in the brain than in the liver or intestinal tract.[72]

From the above-mentioned findings it is evident that the degree of penetration of cadmium into the brain parenchyma across the blood–brain barrier is very limited. In regions lacking this barrier, such as the choroid plexus and hypophysis, the uptake is high. No study seems to have been undertaken as yet to analyze the possible cellular toxicity of cadmium in regions of the brain that are outside the blood–brain barrier. The accumulation of cadmium in the brains of fetuses[72] may occur because the blood–brain barrier is incompletely developed in the immature rat brain.[59]

6.2.2.2. Hemorrhagic Encephalopathy.

In 1967, Gabbiani et al.[37] reported that subcutaneous administration of cadmium chloride to rats and rabbits in doses of 10–20 mg/kg bw resulted in a hemorrhagic encephalopathy. The distribution of the lesions differed between these species, with more pronounced cerebellar changes in rabbits than in rats. In the cerebrum there were hemorrhages in the white matter and in deep regions of the grey matter, with degenerative changes in adjacent nerve cells and glia. The granular

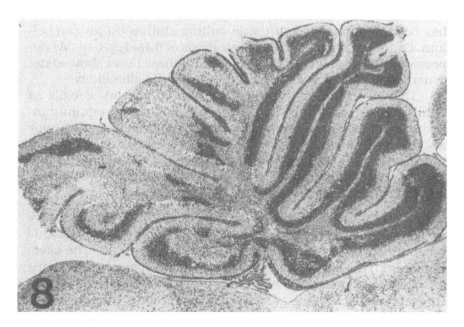

FIG. 8.   Sagittal section through the cerebellum of a mouse injected subcutaneously with a single toxic dose of cadmium on postnatal day 15 and killed after 24 h. Cell necrosis, tissue spaces, and some hemorrhage are seen within the internal granular layer. Reproduced by permission of W.S. Webster and A. A. Valois, *J. Neuropathol. Exp. Neurol.* **40**, 247–257, 1981.

cell layer of the cerebellar cortex was particularly damaged, with cell necrosis[37] (Fig. 8). Recently, Webster and Valois[123] reported a similar encephalopathy in newborn mice. Cadmium chloride was administered subcutaneously in doses of 2–8 mg/kg. bw. After administration on day 1, petechial hemorrhages were diffusely distributed throughout the brain. With increasing age, the lesions were progressively less pronounced, affecting mainly the cerebellum, the olfactory bulbs, and deep layers of the cerebral cortex (Fig. 9). Electron microscopy showed degenerative lesions in endothelial cells of newborn animals 4–6 h after cadmium administration. The endothelial lesions (Fig. 10) were limited to areas with partially differentiated capillaries.[123]

Damage to cerebral blood vessels in fetuses has been described after administration of cadmium ions to pregnant rats. When doses of 2 mg/kg were injected intraperitoneally on day 20 of gestation, the endothelial cells in the caudate nucleus of the fetal brain were found to be vacuolated 1–3 d after cadmium injection. No other ultrastructural abnormalities were seen in the fetal brain.[96]

A direct toxic effect of cadmium on nerve cells within the CNS has been demonstrated in tissue culture studies on rat cerebellum. Cadmium stearate at concentrations of $0.58–1.2 \times 10^{-6} M$ depressed the outgrowth of nerve fibers and these fibers showed degenerative changes such as granules and lytic alterations.[61]

The hemorrhagic lesions in peripheral sensory ganglia of adult animals seem to be a specific feature of acute cadmium intoxication, whereas vascular damage to the immature CNS may also be produced by several other heavy metals. Thus, acute lead intoxication in rats led to hemorrhages and edema of the brain[15,42,85] and chlorides of indium, terbium, thallium, and mercury can induce similar lesions in newborn rats.[39]

6.2.2.3. Influence on Acetylcholine Receptor Binding.

Heavy metals may affect the binding of acetylcholine to the muscarinic receptor in vitro by interaction with sulfhydryl groups. The effect depends on the concentration of the metal, with increased agonist binding at low metal concentrations and inhibition of agonist and then antagonist binding at higher concentrations.[3] Muscarinic receptors in the striatum and cerebral cortex were inhibited in vivo by administration of cadmium in the drinking water to rats. The inhibition was more pronounced in the striatum than in other brain regions, but the reason for this difference is unknown.[53]

6.2.2.4. Changes in Regional Concentrations of Neurotransmitters.

Rats treated with intraperitoneal cadmium

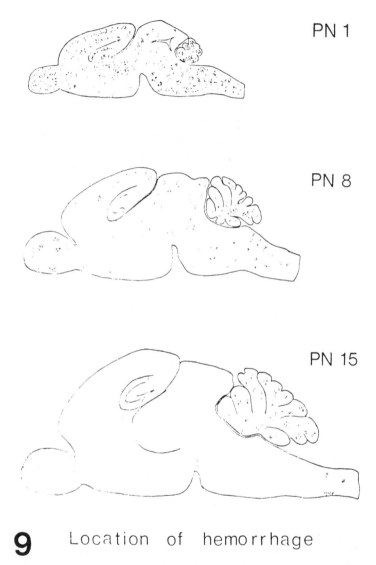

9   Location of hemorrhage

FIG. 9.   Sagittal sections through the brains of mice injected subcutaneously with a single toxic dose of cadmium and killed after 24 h. The distribution of hemorrhages is shown in mice injected on day 1, day 8, or day 15. Each brain ×11. Reproduced by permission of W.S. Webster and A.A. Valois, *J. Neuropathol. Exp. Neurol.* **40,** 247-257, 1981.

chloride injections (0.25–1.0 mg/kg bw) for 45 d developed changes in the levels of various brain biogenic amines. The concentrations of cortical acetylcholine and brainstem serotonin (5-HT) decreased and the cadmium-induced reduction in brainstem 5-HT still persisted even 28 d after withdrawal of treatment.[55]

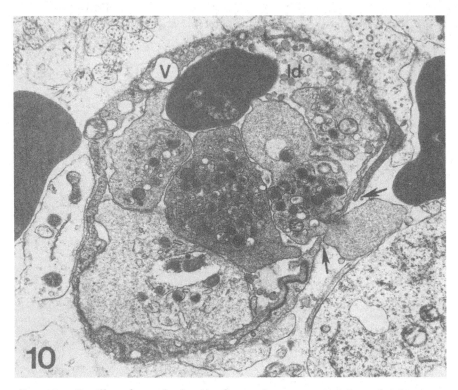

Fig. 10.   Capillary from the brain of a newborn mouse injected subcutaneously with a single toxic dose of cadmium and killed after 8 h. There is
an accumulation of platelets in the vessel. One platelet is plugging a gap
in the endothelial wall (arrow). Note also the accumulation of lipid
droplets(ld), thinning of the capillary wall, and intracellular vacuoles (v).
Reproduced by permission of W.S. Webster and A.A. Valois, *J.
Neuropathol. Exp. Neurol.* 40, 247–257, 1981.

Neonatal rats exposed to low doses of cadmium daily for 30 d
after birth had significantly higher levels of striatal tyrosine
hydroxylase activity and midbrain tryptophan than controls. The
spontaneous locomotor activity was increased in treated rats.[93]

In another study, adult Wistar rats were injected
intravenously with 33–600 μg of Cd^{2+} as cadmium chloride and
the 5-HT and 5-hydroxyindole acetic acid (5-HIAA) levels were
measured in various brain regions including the hypothalamus,
thalamus, mesencephalon, and cortex. The concentration of
5-HIAA increased in all tissues as a function of the cadmium dose.
Serotonin levels were reduced by the lowest cadmium dose, but
were increased in all regions by the other doses of cadmium.[95]

6.2.2.5. Other Effects of Cadmium on the Central Nervous System.

Cadmium may affect the activity of certain enzymes in the brain. Adenylate cyclase activity in the rat cerebrum, cerebellum, and brain stem was inhibited in vitro by micromolar concentrations of cadmium.[22] Monoamine oxidase, which is involved in the catabolism of transmitter monoamines, was also lowered in activity by cadmium in vitro.[74] A comparative study of the effects of cadmium and zinc on cholinesterase activity in the brain showed that cadmium activated cholinesterase, whereas zinc reduced this enzyme after acute intoxication. Following chronic exposure to both metals, the cholinesterase activity was not significantly altered.[87]

Cadmium inhibited rat brain synaptosomal adenosine triphosphatase activity in vitro and was more potent in this respect than aluminum or manganese.[66] All three metals reduced the uptake of choline,[66] catecholamines,[65] γ-amino butyric acid, and glutamic acid[127] by rat brain synaptosomes. Cadmium ions interfere with membrane function in rat brain synaptosomes by inhibiting the methylation of phospholipids. Cadmium has been found to be a much more potent inhibitor in this respect than aluminum or manganese.[126]

Single injections of cadmium acetate potentiated hexobarbital-induced sleep time in rats. The minimal effective dose of cadmium acetate was 2 mg/kg bw. The mechanism by which cadmium alters the drug response in this way is not known.[101]

Rats that received low doses of cadmium in their drinking water before and during pregnancy had pups with a reduced body weight and lower whole body iron than controls. The spontaneous locomotor activity in cadmium-exposed newborn rats was depressed. The reason for this altered behavior is unknown, but it may not represent a toxic effect of cadmium on the CNS since both reduced birth weight and low body iron have an impact on the locomotor activity level.[51]

Visual evoked responses (VER) were studied in rats that had been exposed to cadmium or lead during gestation and lactation. Lead caused long-lasting changes in the VER, whereas cadmium appeared to have no definite effects in this respect unless the animals also had a low body weight or low hematocrit.[19]

Cadmium-induced damage to the testis has been shown to cause hyperplasia of gonadotropic cells in the adenohypophysis of golden hamsters. Corticotropic cells were not affected by the cadmium injection.[41]

7. Future Studies

Experimental studies have thus demonstrated that single large doses of cadmium may produce endothelial cell damage in the nervous system, with hemorrhages and parenchymal destruction. The age and species of the animal greatly influence the distribution of these lesions. Other neurotoxicologic effects in animals are changes in regional concentrations of biogenic amines and inhibition of certain enzymes located in nervous tissue. Most of the experimental studies concern toxic effects of single large doses of cadmium. It would also be of interest to study the effects of long-term oral administration of low doses of cadmium on the nervous system, since this mode of exposure more closely resembles the human situation.

The pathogenesis of endothelial cell damage in ganglia and the immature brain is still obscure. Further experimentation, perhaps with the use of cultured endothelial cells, is necessary to explain why endothelial cell damage is confined to certain regions of the nervous system and why only part of the endothelial cell population in these regions is sensitive to cadmium.

Although the major part of the brain seems to be quite efficiently protected against the penetration of cadmium by the blood–brain barrier, there are certain regions of the CNS that are devoid of this barrier and autoradiographic studies indicate that cadmium may accumulate in such areas. Experimental studies are needed to determine whether the deposition of cadmium in these regions leads to neural cell damage.

Up to now there have been very few reports on neurotoxicologic effects in human beings exposed to cadmium. In view of the numerous observations of toxic effects of cadmium on the nervous system in animal experiments, and considering the increased industrial production of cadmium, it will probably be wise to be observant for neurologic side-effects of cadmium exposure in human beings in the future.

References

1. Adams, R.G. and Crabtree, N. (1961) Anosmia in alkaline battery workers, *J. Industr. Med.* **18,** 216–221.
2. Aoki, A. and Hoffer, A.P. (1978) Reexamination of the lesions in rat testis caused by cadmium. *Biol. Reprod.* **18,** 579–591.
3. Aronstam, R.S., Abood, L.G. and Hoss, W. (1978) Influence of sulfhydryl reagents and heavy metals on the functional state of the

muscarinic acetylcholine receptor in rat brain. *Molec. Pharmacol.* **14**, 575–586.

4. Arvidson, B. (1979) Distribution of intravenously injected protein tracers in peripheral ganglia of adult mice. *Exp. Neurol.* **63**, 388–410.

5. Arvidson, B. (1980) Regional differences in severity of cadmium induced lesions in the peripheral nervous system in mice. *Acta Neuropathol.* **49**, 213–224.

6. Arvidson, B. (1981) Is cadmium toxic to the nervous system? *Trends Neurosci.* **4**, 11–13.

7. Arvidson, B. and Tjälve, H. (1983) Distribution of [109]Cd in the nervous system of rats after intravenous injection. *Neurotoxicol. Neurobiol. in press.*

8. Aylett, B.J. (1979) The chemistry and bioinorganic chemistry of cadmium. In: M. Webb (Ed.), *The Chemistry, Biochemistry and Biology of Cadmium*, Elsevier, Amsterdam, pp. 1–44.

9. Baader, E.W. (1952) Chronic cadmium poisoning. *Ind. Med. Surg.* **21**, 427–430.

10. Berlin, M. and Ullberg, S. (1963) The fate of [109]Cd in the mouse. *Arch. Environm. Health* **7**, 686–693.

11. Betz. A.L., Firth, J.A. and Goldstein, G.W. (1980) Polarity of the bloodbrain barrier: Distribution of enzymes between the luminal and antiluminal membranes of brain capillary endothelial cells. *Brain Res.* **192**, 17–28.

12. Björklund, H., Hoffer, B., Olson, L. and Seiger, Å. (1981) Differential morphological changes in sympathetic nerve fibers elicited by lead, cadmium and mercury. *Environm. Res.* **26**, 69–80.

13. Buhler, D.R., Wright, D.C., Smith, K.L. and Tinsley, I.J. (1981) Cadmium absorption and tissue distribution in rats provided low concentrations of cadmium in food or drinking water. *J. Toxicol. Environm. Health* **8**, 185–197.

14. Carrol, R.E. (1966) The relationship of cadmium in the air to cardiovascular disease death rates. *J. Amer. Med. Assn.* **198**, 267–270.

15. Clasen, A.A., Hartman, J.J., Starr, A.J., Coogan, P.S., Pandolfi, S., Laing, I., Becker, R. and Hass, G.M. (1974) Electron microscopic and chemical studies of lead encephalopathy: a comparative study of the human and experimental disease. *Am. J. Pathol.* **74**, 215–233.

16. Cole, G.M. and Baer, L.S. (1944) Food poisoning from cadmium. *Nav. Med. Bull.* **43**, 398–403.

17. Cooper, G.P. and Steinberg, O. (1977) Effects of cadmium and lead on adrenergic neuromuscular transmission in the rabbit. *Am. J. Physiol.* **232**, C128–C131.

18. Cooper, G.P., Kober, T.E. and Manalis, R.S. (1980) Differences in the effects of cadmium and lead on synaptic transmission. *Soc. Neurosci. Abstr.* **6**, 176.

19. Cooper, G.P., Fox, D.A., Howell, W.E., Laurie, R.D., Tsang, W. and Lewkowski, J.P. (1980) Visual evoked responses in rats exposed to heavy metals. In: W.H. Merigan and B. Weiss (Eds.), *Neurotoxicity of the Visual System*, Raven Press, New York, pp. 203–218.

20. Davis, J.S., Flynn, F.V. and Platt, H.S. (1968) The characterization of urine protein by gel filtration. *Clin. Chim. Acta* **21**, 357–362.

21. Decker, C.F., Byerrum, R.U. and Hoppert, C.A. (1957) A study of the distribution and retention of cadmium-115 in the albino rat. *Arch. Biochem.* **66**, 140–144.

22. Ewers, U. and Erbe, R. (1980) Effects of lead, cadmium and mercury on brain adenylate cyclase. *Toxicol.* **16**, 227–237.

23. Fassett, D.W. (1972) Cadmium. In: D.H.K. Lee (Ed.), *Metallic Contaminants and Human Health*, Academic Press, New York, pp. 97–126.

24. Fassett, D.W. (1975) Cadmium: Biological effects and occurrence in the environment. *Ann. Rev. Pharmacol.* **15**, 425–435.

25. Ferm, V.H. and Carpenter, S.J. (1968) The relationship of cadmium and zinc in experimental mammalian teratogenesis. *Lab. Invest.* **18**, 429–432.

26. Flanagan, P.R., McLellan, J.S., Haist, J., Cherian, M.G., Chemberlain, M.J. and Valberg, L.S. (1978) Increased dietary cadmium absorption in mice and human subjects with iron deficiency. *Gastroenterol.* **74**, 841–846.

27. Fleischer, M., Sarofim, A.F., Fassett, D.W., Hammond, P., Shacklette, H.T., Nisbet, I.C.T. and Epstein, S. (1974) Environmental impact of cadmium: A review by the panel on hazardous trace substances. *Environm. Health Persp.* **7**, 253–323.

28. Forshaw, P.J. (1977) The inhibitory effect of cadmium on neuromuscular transmission in the rat. *Eur. J. Pharmacol.* **42**, 371–377.

29. Foster, C.L. and Cameron, E. (1963) Observations on the histological effects of sub-lethal doses of cadmium chloride in the rabbit. II. The effect on the kidney cortex. *J. Anat.* **97**, 281–285.

30. Friberg, L. (1948) Proteinuria and emphysema among workers exposed to cadmium and nickel dust in a storage battery plant. *Proc. Int. Congr. Ind. Med.* **9**, 641–645.

31. Friberg, L. (1950) Health hazards in the manufacture of alkaline accumulators with special reference to chronic cadmium poisoning. *Acta Med. Scand.* **138**, Suppl 240.

32. Friberg, L. (1957) Deposition and distribution of cadmium in man in chronic poisoning *Arch. Ind. Health* **16**, 27–29.

33. Friberg, L., Piscator, M., Nordberg, G.F. and Kjellström, T. (1974) *Cadmium in the Environment.* CRC Press Inc., Cleveland.

34. Friberg, L., Kjellström, T., Nordberg, G.F. and Piscator, M. (1979) Cadmium. In: L. Friberg, G.F. Nordberg, and V.B. Vouk (Eds.), *Handbook on the Toxicology of Metals*, Elsevier, Amsterdam, pp. 355–382.

35. Gabbiani, G. (1966) Action of cadmium chloride on sensory ganglia. *Experientia* **22**, 261–262.

36. Gabbiani, G. (1969) Endothelial microvilli in the vessels of the rat gasserian ganglion and testis. *Z. Zellforsch.* **97**, 111–117.

37. Gabbiani, G., Baic, D. and Déziel, C. (1967) Toxicity of cadmium for the central nervous system. *Exp. Neurol.* **18**, 154–160.

38. Gabbiani, G., Baic, D. and Déziel, C. (1967) Studies on tolerance and ionic antagonism for cadmium and mercury. *Can. J. Physiol. Pharmacol.* **45**, 443–450.

39. Gabbiani, G., Gregory, A. and Baic, D. (1967) Cadmium-induced selective lesions of sensory ganglia. *J. Neuropath. Exp. Neurol.* **26**, 498–506.

40. Gabbiani, G., Badonnel, M.-C., Mathewson, S.M. and Ryan, G.B. (1974) Acute cadmium intoxication. Early selective lesions of endothelial clefts. *Lab. Invest.* **30**, 686–695.

41. Girod, C. and Dubois, M.P. (1976) Influence du chlorure de cadmium sur l'axe hypophyso-testiculaire du hamster doré. *Ann. Endocrinol. (Paris)* **37**, 273–274.

42. Goldstein, G.W., Asbury, A.K. and Diamond, I. (1974) Pathogenesis of lead encepthalopathy—uptake of lead and reaction of brain capillaries. *Arch. Neurol.* **31**, 382–389.

43. Gunn, S.A., Gould, T.C. and Anderson, W.A.D. (1963) Cadmium-induced interstitial cell tumours in rats and mice and their prevention by zinc. *J. Nat. Cancer Inst.* **31**, 745–749.

44. Gunn, S.A., Gould, T.C. and Anderson, W.A.D. (1967) Specific response of mesenchymal tissue to cancerogenesis by cadmium. *Arch. Pathol.* **83**, 493–496.

45. Haddow, A., Dukes, C.E. and Mitchley, B.C.V. (1961) Carcinogeneity of iron preparations and metal–carbohydrate complexes. *Rep. Br. Emp. Cancer Campn.* **39**, 74–78.

46. Haddow, A., Roe, F.J.C., Dukes, C.E. and Mitchley, B.C.V. (1964) Cadmium neoplasia: Sarcomata of the site of injection of cadmium sulphate in rats and mice. *Br. J. Cancer* **18**, 667–673.

47. Hagino, N. (1957) About investigations on Itai-itai disease. *J. Toyama Med. Assoc. Suppl.*, Dec. 21.

48. Hagino, N. (1973) Itai-itai disease and vitamin D. *Dig. Sci. Labour* **28**, 32–36.

49. Hamada, T., Iwamasa, T., Tsuru, T., Sato, K. and Takeuchi, T. (1980) Mechanism of glycogenosome formation in axons of cadmium-induced neuropathy. Ultrastructural and biochemical studies. *Neurotoxicol.* **2**, 33–41.

50. Harrison, J.F., Lunt, G.S., Scott, P. and Blainey, J.D. (1968) Urinary lysozyme, ribonuclease and low-molecular weight protein in renal disease. *Lancet* **1**, 371–380.

51. Hastings, F., Choudhury, H., Petering, H.G. and Cooper, G.P. (1978) Behavioral and biochemical effects of low-level prenatal cadmium exposure in rats. *Bull. Environ. Contam. Toxicol.* **20**, 96–101.

52. Hayashi, H. and Takayama, K. (1978) Inhibitory effects of cadmium on the release of acetylcholine from cardiac nerve terminals. *Jap. J. Physiol.* **28**, 333–345.

53. Hedlund, B., Gamarra, M. and Bartfai, T. (1979) Inhibition of striatal muscarinic receptors in vivo by cadmium. *Brain Res.* **168**, 216–218.

54. Hoffman, E.O., Cook, J.A., DiLuzio, N.R. and Coover, J.A. (1975) The effects of acute cadmium administration in the liver and kidney of the rat. Light and electron microscopic studies. *Lab. Invest.* **32,** 655–673.

55. Hrdina, P.D., Peters, D.A.V. and Singhal, R.L. (1976) Effects of chronic exposure to cadmium, lead and mercury on brain biogenic amines in the rat. *Res. Comm. Chem. Pathol. Pharmacol.* **15,** 483–493.

56. Ishizaki, A., Fukushima, M. and Sakamoto, M. (1970) Distribution of Cd in biological materials. II. Cadmium and zinc contents of foodstuffs. *Jap. J. Hyg.* **25,** 207–211.

57. Jacobs, J.M. (1977) Penetration of systemically injected horseradish peroxidase into ganglia and nerves of the autonomic nervous system. *J. Neurocytol.* **6,** 607–618.

58. Jacobs, J.M., MacFarlane, R.M. and Cavanagh, J.B. (1976) Vascular leakage in the dorsal root ganglia of the rat, studied with horseradish peroxidase. *J. Neurol. Sci.* **29,** 95–107.

59. Johanson, C.E. (1980) Permeability and vascularity of the developing brain: cerebellum vs cerebral cortex. *Brain Res.* **190,** 3–16.

60. Kägi, J.H.R. and Vallee, B.L. (1960) Metallothionein: a cadmium and zinc-containing protein from equine renal cortex. *J. Biol. Chem.* **235,** 3460–3465.

61. Kasuya, M., Sugawara, N. and Okada, A. (1974) Toxic effect of cadmium stearate on rat cerebellum in culture. *Bull. Environ. Contam. Toxicol.* **12,** 535–540.

62. Kawai, K., Fukuda, K. and Kimura, M. (1976) In: *Effects and Dose–Response Relationships of Toxic Metals.* G.F. Nordberg (Ed.), Elsevier, Amsterdam, pp. 343–370.

63. Kipling, M.D. and Waterhouse J.A.H. (1967) Cadmium and prostatic carcinoma. *Lancet* **1,** 730–733.

64. Kjellström, T., Friberg, L. and Rahnster, B. (1979) Mortality and cancer morbidity among cadmium-exposed workers. *Environ. Health. Persp.* **28,** 199–205.

65. Lai, J.C.K., Guest, J.F., Lim, L. and Davison, A.N. (1978) The effects of transition-metal ions on rat brain synaptosomal amine-uptake systems. *Biochem. Soc. Trans.* **6,** 1010–1012.

66. Lai, J.C.K., Guest, J.F., Leung, T.K.C., Lim, L. and Davison, A.N. (1980) The effects of cadmium, manganese and aluminum on sodium-potassium-activated and magnesium-activated adenosine triphosphatase activity and choline uptake in rat brain synaptosomes. *Biochem. Pharmacol.* **29,** 141–146.

67. Lane, R.E. and Campbell, A.C.P. (1954) Fatal emphysema in two men making a copper cadmium alloy, *Brit. J. Industr. Med.* **11,** 118–122.

68. Larsson, S.-E, and Piscator, M. (1971) Effect of cadmium on skeletal tissue in normal and calcium-deficient rats *Isr. J. Med. Sci.* **7,** 495–502.

69. Lauwerys, R. (1979) Cadmium in man. In: M. Webb (Ed.) *The Chemistry, Biochemistry and Biology of Cadmium,* Elsevier, Amsterdam, pp. 449–452.

70. Lemen, R.A., Lee, J.S., Wagoner, J.K. and Blejer, H.P. (1976) Cancer mortality among cadmium production workers. *Ann. NY Acad. Sci.* **271**, 273–276.

71. Lewis, G.P., Jusko, W.J. and Coughlin, L.L. (1972) Cadmium accumulation in man: influence of smoking, occupation, alcoholic habit, and disease. *J. Chronic Dis.* **25**, 717–723.

72. Lucis, O.J., Lucis, R. and Shaikh, Z.A. (1972) Cadmium and zinc in pregnancy and lactation. *Arch. Environ. Health.* **25**, 14–22.

73. Lucis, O.J., Lynk, M.E. and Lucis, R. (1969) Turnover of cadmium 109 in rats. *Arch. Environ. Health* **18**, 307–312.

74. Magour, S., Cumpelik, O. and Paulus, M. (1979) Effect of cadmium and copper on monoamine oxidase type A and B in brain liver mitochondria. *J. Clin. Chem. Clin. Biochem.* **17**, 777–780.

75. Margoshes, M. and Vallee, B.C. (1957) A cadmium protein from equine kidney cortex. *J. Am. Chem. Soc.* **79**, 4813–4814.

76. Marmé, W. (1867) Über die giftige Wirkung und den Nachweis einiger Cadmium-Verbladungen. *Ztschr. Rationel. Med.* **29**, 125–130.

77. Moritsugi, M. and Kobayashi, J. (1964) Study on trace metals in biomaterials. II. Cadmium content in polished rice. *Ber. Ohara Inst. Landwirtsch. Biol. Okayama Univ.* **12**, 145–149.

78. Nordberg, G.F. (1972) Cadmium metabolism and toxicity. *Environ. Physiol. Biochem.* **2**, 7–36.

79. Nordberg, G.F. and Nishiyama, K. (1972) Whole-body and hair retention of cadmium in mice. Arch. Environ. Health **24**, 209–214.

80. Nordberg, M. and Nordberg, G.F. (1975) Distribution of metallothionein-bound cadmium and cadmium chloride in mice: preliminary studies. *Environ. Health Persp.* **12**, 103–108.

81. Olsson, Y. (1971) Studies on vascular permeability in peripheral nerves. IV. Distribution of intravenously injected protein tracers in the peripheral nervous system of various species. *Acta Neuropathol.* **16**, 103–116.

82. Parizek, J. (1956) Effect of cadmium salts on testicular tissue. *Nature* **177**, 1036–1037.

83. Parizek, J. (1957) The destructuve effect of cadmium on testicular tissue and its prevention by zinc. *J. Endocrinol.* **15**, 56–59.

84. Paterson, J.C. (1947) Studies on the toxicity of inhaled cadmium. III. The pathology of cadmium smoke poisoning in man and in experimental animals. *J. Ind. Hyg. Toxicol.* **29**, 294–302.

85. Pentschew, A. and Garro, F. (1966) Lead encephalo-myelopathy of the suckling rat and its implications on the porphyrinopathic nervous system. *Acta Neuropathol.* **6**, 266–278.

86. Perry, H.M. and Erlanger, M.W. (1971) Hypertension and tissue metal levels after intraperitoneal cadmium, mercury, and zinc. *Am. J. Physiol.* **220**, 808–811.

87. Pham-Huu-Chanh and Plancade, Y. (1971) Etude comparée des effets du zinc et du cadmium sur les activites cholinesterasiques tissulaires du rat. *Biochem. Pharmacol.* **20**, 729–736.

88. Piscator, M. (1981) Role of cadmium in carcinogenesis with special reference to cancer of the prostate. *Environ. Health Persp.* **40**, 107–120.

89. Potts, C.L. (1965) Cadmium proteinuria—the health of battery workers exposed to cadmium oxide dust. *Ann. Occup. Hyg.* **8**, 55–62.

90. Pringle, B.H., Hissong, D.E., Katz, E.L. and Mulawka, S.T. (1968) Trace metal accumulation by estuarine mollusks. *J. Sanit. Eng. Div. Proc. Am. Soc. Civ. Eng.* **94**, 455–463.

91. Prodan, L. (1932) Cadmium poisoning: I. The history of cadmium poisoning and uses of cadmium. *J. Ind. Hyg.* **14**, 132–155.

92. Rahola, T., Aaran, R.-K. and Miettinen, J.K. (1971) Half-time studies of mercury and cadmium by whole body counting. IAEA Symposium on the Assessment of Radioactive Contamination in Man, IAEA, Vienna.

93. Rastogi, R.B., Merali, Z. and Singhal, R.L. (1977) Cadmium alters behaviour and the biosynthetic capacity for cathecolamines and serotonin in neonatal rat brain. *J. Neurochem.* **28**, 789–794.

94. Reynolds, C.V. and Reynolds, E.B. (1971) Cadmium in crabs and crabmeat. *J. Assoc. Public Anal.* **9**, 112–115.

95. Ribas-Ozonas, B., Ochoa Estomba, M.C. and Santos-Ruiz, A. (1974) Activation of serotonin and 5-hydroxyindolacetic acid in brain structures after application of cadmium. In: W.G. Hoekstra, J.W. Suttie, H.E. Ganther and W. Mertz (Eds.) *International Symposium on Trace Element Metabolism in Animals—2*, University Park press, Baltimore, pp. 476–478.

96. Rohrer, S.R., Shaw, S.M. and Lamar, C.H. (1978) Cadmium induced endothelial cell alteration in the fetal brain from prenatal exposure. *Acta Neuropath.* **44**, 147–149.

97. Romanul, F.C., Bannister, R.G. (1962) Localized areas of high alkaline phosphatase activity in endothelium of arteries. *Nature* **195**, 611–612.

98. Samarawickrama, G.P. (1979) Biological effects of cadmium in mammals. In: M. Webb (Ed.) *The Chemistry, Biochemistry and Biology of Cadmium.* Elsevier, Amsterdam, pp. 341–422.

99. Sato, K., Iwamasa, T., Tsuru, T. and Takeuchi, T. (1978) An ultrastructural study of chronic cadmium-induced neuropathy. *Acta Neuropathol.* **41**, 185–190.

100. Schlaepfer, W.W. (1971) Sequential study of endothelial changes in acute cadmium intoxication. *Lab Invest.* **25**, 556–564.

101. Schnell, R.C., Prosser, T.D. and Miya, T.S. (1974) Cadmium-induced potentiation of hexobarbital sleep time in rats. *Experientia* **30**, 528–529.

102. Schroeder, H.A. (1964) Cadmium hypertension in rats. *Am. J. Physiol.* **207**, 62–68.

103. Schroeder, H.A. (1965) Cadmium as a factor in hypertension. *J. Chronic Dis.* **18**, 647–653.

104. Schroeder, H.A. and Mitchener, M. (1971) Toxic effects of trace elements on the reproduction of mice and rats. *Arch. Environ. Health* **23**, 102–106.

105. Schroeder, H.A., Nason, A.P. and Balassa, J.J. (1967) Trace metals in rat tissues as influenced by calcium in water. *J. Nutr.* **93**, 331–339.

106. Sharma, R.P. and Obersteiner, E.J. (1981) Metals and neurotoxic effects: cytotoxicity of selected metallic compounds on chick ganglia cultures. *J. Comp. Path.* **91**, 235–244.

107. Smith, J.C., Kench, J.E. and Smith, J.P. (1957) Chemical and histological post-mortem studies on a workman exposed for many years to cadmium oxide fume. *Brit. J. Industr. Med.* **14**, 246–249.

108. Smithells, C.J. (1976) *Metals Reference Handbook*, 5th ed, Butterworths, London.

109. Sovet (1858) Empoisonnement par une poudre à ecurer l'argenterie. *Presse Méd. Belge* **10**, 69–70.

110. Suzuki, S., Taguchi, T. and Yokohashi, G. (1969) Dietary factors influencing upon the retention rate of orally administered $^{115m}CdCl_2$ in mice with special reference to calcium and protein concentrations in diet. *Ind. Health* **7**, 155–160.

111. Szadkowski, D., Schultze, H., Schaller, K.H. and Lehnert, G. (1969) Zur ökologischen Bedeutung des Schwermetallgehaltes von Zigaretten. *Arch. Hyg. Bakteriol.* **153**, 1–8.

112. Tischner, K.H. (1980) Cadmium. In: P.S. Spencer and H.H. Schaumburg (Eds.) *Experimental and Clinical Neurotoxicology*, Williams and Wilkins, Baltimore, pp. 348–355.

113. Tischner, K.H. and Schröder, J.M. (1972) The effects of cadmium chloride on organotypic cultures of rat sensory ganglia. *J. Neurol. Sci.* **16**, 383–399.

114. Toda, N. (1973) Influence of cadmium ions on the transmembrane potential and contractility of isolated rabbit left atria. *J. Pharmacol. Exp. Therap.* **186**, 60–66.

115. Toda, N. (1976) Neuromuscular blocking action of cadmium and manganese in isolated frog striated muscle. *Eur. J. Pharmacol.* **40**, 67–95.

116. Todd, A.S. (1964) Some topographical observations on fibrinolysis. *J. Clin. Pathol.* **17**, 324–327.

117. Vallee, B.L. and Ulmer, D.D. (1972) Biochemical effects of mercury, cadmium and lead. *Ann. Rev. Biochem.* **41**, 91–96.

118. Vorobjeva, R.S. (1957a) On occupational lung disease in prolonged action of aerosol of cadmium oxide. *Arch. Patologii* **8**, 25–30.

119. Vorobjeva, R.S. (1975b) Investigations of the nervous system function in workers exposed to cadmium oxide. *In. Neuropat. Psikhiat.* **57**, 385–393.

120. Walsh, J.J. and Burch, G.E. (1959) The rate of disappearance from plasma and subsequent distribution of radiocadmium (^{115m}Cd) in normal dogs. *J. Lab. Clin. Med.* **54**, 59–65.

121. Webb, M. (1972) Binding of cadmium ions by rat liver and kidney. *Biochem. Pharmacol.* **21**, 2751–2765.
122. Webb, M. (1975) Cadmium. *Br. Med. Bull.* **31**, 246–250.
123. Webster, W.S. and Valois, A.A. (1981) The toxic effects of cadmium on the neonatal mouse CNS. *J. Neuropath. Exp. Neurol.* **40**, 247–257.
124. Weeks, M.E. and Leicester, H.M. (1968) Discovery of the Elements. *J. Chem. Ed.*, Easton, pp. 502–509.
125. Wisniewska-Knypl, I.M., Jablonska, J. and Myslak, Z. (1971) Binding of cadmium on metallothionein in man: an analysis of a fatal poisoning by cadmium iodide. *Arch. Toxikol.* **28**, 46–53.
126. Wong, P.C.L. and Lim, L. (1981) The effects of aluminium, manganese and cadmium chloride on the methylation of phospholipids in the rat brain synaptosomal membrane. *Biochem. Phamacol.* **30**, 1704–1705.
127. Wong, P.C.L., Lai, J.C.K., Lim, L. and Davison, A.N. (1981) Selective inhibition of L-glutamate and gammaaminobutyrate transport in nerve ending particles by aluminium, manganese and cadmium chloride. *J. Inorg. Biochem.* **14**, 253–260.
128. Wood, R.W. (1978) Stimulus properties of inhaled substances. *Env. Health Persp.* **26**, 69–76.

Chapter 3

Cadmium and Teratogenesis of the Central Nervous System

Louis J. Pierro

1. Introduction

Effects of cadmium on the developing embryo or fetus have been studied by many investigators since the report by Parizek in 1964[35] that a single maternal injection of cadmium at gestation days 17–21 resulted in a destruction of the fetal portion of the placenta and fetal death. The first published reports of cadmium-induced teratogenesis described studies with the hamster[14,15]; cadmium-induced teratogenicity was subsequently reported for the rat,[1,5] and for the mouse.[30,31,37–39] Using a different experimental approach, Klein et al.[29] reported abnormal development of head-fold stage rat embryos cultured in vitro on serum taken from rats injected with cadmium.

Malformations induced by cadmium treatment of pregnant hamsters include defects of the face and upper jaw, exencephaly, anophthalmia, rib fusions, and limb defects.[13–16,34] Craniofacial defects were produced more frequently by treatment early on day 8 of gestation; limb defects by treatment on day 9. In the mouse also, craniofacial defects were observed after treatment at 8 d

Louis J. Pierro: Department of Animal Genetics, University of Connecticut, Storrs, Connecticut.

gestational age,[37–39] and limb defects after treatment at 9 d gestation.[30,31] Common abnormalities observed in rat fetuses challenged by maternal treatment with cadmium on day 9 (approximately equivalent to 8 d gestation in the mouse[43]) included eye defects and hydrocephaly.[1] The frequency of eye defects is low after treatment at 10 d; hydrocephaly was not observed after treatment on day 10, but reappeared after treatment on day 11. Treatment at day 13 and later produced jaw defects, cleft palate, and hypoplastic lungs.[5] More recently, Takeuchi et al. [44] have reported craniofacial defects following treatment of MPI albino rats with cadmium sulfate (3–5 mg/kg) on day 9 of gestation. Treatment on day 10 produced digital anomalies of hind limbs, anal atresia, and tail anomalies.

2. Effects on the Neural Tube and Its Derivatives

In the hamster, maternal treatment (iv) with cadmium (2 mg/kg) at 10 AM on day 8 resulted in eye defects in 25% of the fetuses and neural tube defects in 44%.[13] A marked reduction in the frequency of eye and neural tube defects was observed following treatment at 8 PM on day 8. In mice of the C57BL/10ChPr strain, treatment (ip) with cadmium (0.02 mM/kg) at 8–10 AM on day 8 of gestation resulted in 88% eye defects and 79% neural tube defects.[39] (Note that data in this paper were reported on the basis of gestational age rather than day of gestation.) In most cases, eye and neural tube defects were observed in the same fetuses; only 4.6% showed isolated exencephaly in this treatment group. Treatment 24 h later resulted in only 18.4% of the fetuses showing neural tube defects, and the frequency of eye defects (14.1%) was close to the level of spontaneous defects characteristic of the strain. Exencephaly was the most common neural tube abnormality, but encephalocele was observed upon external examination of both hamster and mouse fetuses. Internal examination based on both histological and Wilson sections revealed some cases of hydrocephalus interna in cadmium-treated mouse fetuses.

Exencephaly and associated eye defects were also observed in rat embryos surviving culture in vitro for 48 h on serum from cadmium-treated rats.[29] Since the embryos were explanted at head-fold stage, the culture test period overlapped with the gestational ages challenged in the *in utero* studies reported by Barr.[1] Exencephaly was not seen in the *in utero* studies, but eye defects and hydrocephalus were. More recently, Takeuchi et al.[44] reported exencephaly and encephalocele, as well as eye and facial defects, following treatment of MPI albino rats with cadmium sulfate (3–5 mg/kg) on day 9 of gestation. Treatment on day 8 produced only eye defects. Another expression of interference with neural tube development has been reported[36] in studies with Holtzman rats that received up to 2 mg/kg bw cadmium acetate on gestation days 8, 10, or 12. Defects were described as doming of the calvarium of the fetal head and attenuation of the neck area. Examination of histological sections from fetuses with domed skulls indicated an increase in the dorso-ventral dimension of the brain, accompanied by a shortening of the antero-posterior axis; the lateral ventricles were approximately one-third longer along their dorso-ventral axis than expected.

For the most part, studies of the effects of cadmium on the developing nervous system have focused attention on essentially major defects that can be readily detected using the classical methodology of teratology. In virtually all such studies, some embryos fail to show any apparent defects with these approaches. These embryos are operationally classified as defect-free, although it is generally recognized that examination using other techniques might in fact reveal some developmental aberrations. Behavioral effects, for example, have been demonstrated in studies of chemical agents also known under certain conditions to produce gross malformations of the central nervous system in rats and mice.[3,40]

To date, relatively few attempts to measure behavioral effects of cadmium treatment during the prenatal period have been published. Two papers published in 1978 have demonstrated depression of spontaneous locomotor activity (SLA) in male progeny born to females given cadmium at 17.2 μg/mL in their drinking water during a pre-mating period (90 d) as well as during gestation.[9,23] In contrast, cadmium at 34.4 μg/mL appeared to be associated with an increase in SLA.[9] Doses of 4.3 and 8.6 μg/mL were without effect. The cellular basis of these effects was not determined.

Tsujii and Hoshishima[47] have performed a comprehensive series of behavioral tests in order to monitor development of reflex responses in mice born to mothers that received 11 daily intraperitoneal injections (10 ng/mouse) during gestation (3 consecutive days followed by 8 alternate days). These authors concluded that cadmium treatment influenced development of reflex responses, either accelerating or retarding the process depending on the specific test administered. Tsujii and Hoshishima also re-

ported a significantly increased number of errors in a maze learning test for the progeny of the cadmium-treated females. The cellular basis of these effects was not investigated.

Recently, Webster and Valois[52] have published a report which
begins to provide a basis for relating behavioral anomalies produced by cadmium treatment to cellular and tissue alterations in
the immature brain. Outbred QS mice received single subcutaneous injections of cadmium chloride on postnatal days 1, 8, 15, or
22. Petechial hemorrhages, edema, and cellular pycnosis were detected throughout the brain, 24 h after cadmium exposure on day
1. Exposure on days 8 or 15 produced similar, but more restricted,
damage. Treatment on day 22 was apparently without effect. Simple behavioral tests (exploratory behavior in an open field, balan-
cing on a horizontal rod) indicated that animals that survived 8
weeks after treatment on days 1 or 8 were hypoactive and lacked
coordination. Brains of animals treated on day 1 were very small,
the cerebral cortex was very thin, and the lateral ventricles dilated.
Brains of animals treated on day 8 showed considerable destruction of the forebrain. The lateral vesicles were massively dilated,
the cerebral cortex was very thin, and the basal ganglia were almost completely missing. In animals treated on either day, the cerebellum was small, but appeared normal. Animals surviving treatment on day 15 appeared to be behaviorally normal. Some
thinning of the cerebral cortex and reduction in size of the cerebellum were apparent.

3. Ontogenesis of Cadmium-Induced Neural Tube Defects

The sequence of events occurring during closure of the neural
tube in rodent embryos includes (a) the appearance of a distinct
boundary between the neural plate and surface ectoderm; (b) flat-
tening of cells in the boundary region, accompanied by the appearance of extensive surface alterations (ruffles in the mouse;
blebs in the hamster); (c) contact of neural folds first between the
flattened cell zone and subsequently involving surface ectoderm;
and (d) formation of neural crest by cells subjacent to the zone of
morphological alterations.[50] Neural folds converge initially at the
level of the posterior hindbrain–anterior trunk in embryos of 4–5
somites. By this stage, cranial flexure has begun and the optic sul-
cus is present; the allantois extends about half-way to the chorion.
Within approximately 0.5 d of gestation the neural folds have fused

at the site of initial contact and the contact-fusion process is spreading both cranially and caudally. During the same period, neural fold contact and fusion take place in the prosencephalon at both the extreme anterior and posterior ends, and spread caudally from each site. The fusion process is completed relatively quickly over the prosencephalon, more slowly over the mesencephalon. The anterior neuropore (last opening into the anterior neural tube) is located at the level of the hindbrain in rodent embryos, with the closure process converging in a caudal direction from the mesencephalon and in a cranial direction from the posterior myelencephalon. Anterior and posterior neuropores are closed during the 9th d of gestation in the mouse and during the 10th d in the hamster. In the rat embryo, closure of the neural tube takes place between the four and 14 somite stages, and closely resembles that reported for the hamster and mouse embryo.[33] Observations in the rat embryo suggest particular importance for cranial flexure, elongation of the open neural fold area, microfilament dynamics, and neural crest migration.

Morriss and New[33] have pointed out that contact at the forebrain apposition point follows a rapid narrowing of the V-shaped opening in the neural folds that seems to be associated with development of a pronounced cranial flexure. A recent study by Goodrum and Jacobson[20] using chick embryos suggests that cranial flexure is accompanied by differential growth of the mesencephalic roof relative to the floor. Bending of the mesencephalon can occur in heads explanted onto the chorioallantoic membrane or into an in vitro system. Goodrum and Jacobson [20] point out that there is also differential growth of the prosencephalic roof relative to the floor, even though the prosencephalon does not bend. Presumably the growth is taken up in the various changes in shape accompanying morphogenesis within the prosencephalon. The authors conclude "that while differential growth may be necessary for flexure, the growth *per se* does not effect the change in shape. Rather the change of shape must be directed by many factors that decide the disposition in space of the products of growth."

Morriss and New[33] also suggest that closure at the level of the broad cranial neural folds (mesen-, meten-, myelencephalon) involves stretching forces that may depend on elongation of the notochord. Jacobson and Gordon[27] had previously suggested that notochord expansion might be involved in aspects of neural plate and neural fold formation in amphibian embryos. The stretching forces, whatever their basis, are assumed to be aided by the appearance of a microfilament-free region at the lateral edge of the neuroepithelium, migration of neural crest cells, and normal cell death.

At the present time little is known about the basis for the failure of the neural tube to close normally in cadmium-treated embryos. The first reported sign of an embryonic response to cadmium treatment was the presence at 12 h of intracellular inclusions that were membrane-bound, variable in size and density, and darkly staining in Bouin's fixed preparations stained with hematoxylin and eosin.[51] Embryos collected at 12 h after treatment on gestation day 7 (egg cylinder stage) showed inclusion bodies in the embryonic ectoderm. In the case of embryos collected after treatment on gestation day 8, inclusion bodies were found in neuroepithelial cells of both open and closed parts of the neural tube, and in cephalic neural crest cells. Embryos collected after treatment on gestation days 9 or 10 showed numerous inclusion bodies in cells of the neural tube and neural crest, where present, and to some extent in cells of the limb buds and somites. Inclusion bodies were still present in some cells of embryos collected at 24 h in groups treated on gestation days 7 and 8, but were not present at 48 h. In embryos treated on gestation day 9 or 10, inclusion bodies and cell fragments were found within prominent extracellular debris; cell inclusions were no longer present at 48 h.

Unpublished observations from this laboratory confirm the presence of densely staining bodies (hematoxylin–PAS or hematoxylin–eosin) in the neuroepithelium of cadmium-treated mouse embryos, in some cases as early as 8 h after treatment. Presumably these correspond to the intracellular inclusions demonstrated with transmission electron microscopy by Webster and Messerle,[51] although they have been studied only at the light microscope level (paraffin sections cut at 7 μm). Densely stained inclusion bodies have also been seen among cephalic neural crest cells, and in one embryo (48 h post-treatment at the beginning of gestation day 8) among cells found within the otic placode. Except for this peculiar embryo, inclusion bodies were not conspicuous in embryos collected 48 h post-treatment unless the embryos appeared moribund (disorganization of heart and other body tissues, extensive cell debris in gut, and so on).

Webster and Messerle[51] reported that in the C57BL/6J strain used in their studies, the neural tube was closed in embryos collected on gestation day 9 except at the anterior and posterior neuropore. The neural tube of control embryos collected on gestation day 10 was completely closed. In embryos collected on gestation day 10, following treatment on either gestation day 7 or 8, the neural tube was either partially open in the cranial region (exencephaly) or closed. In neither case did the neuroepithelium show any sign of cell damage. In the case of embryos collected 48 h after cadmium treatment on gestation days 9 or 10, the neural tube of most seemed normal. Examples were found in which small sections 10–20 μm of the trunk neural tube were open dorsally, but no signs of cellular damage were apparent. The implication from these studies is that the cellular effects of cadmium are relatively short-lived. Their relationship to failure of neural tube closure requires further study.

Studies in this laboratory on mouse embryos collected at 24 h intervals subsequent to treatment at the beginning of gestation day 8 have resulted in a number of observations that have to be considered in any explanation of the interference by cadmium with neural tube closure. (1) Interference with closure of the neural tube in the trunk region is relatively rare; most frequently, neural tube closure is relatively normal throughout the trunk and into the hindbrain to about the level of the otocyst. (2) Interference with neural tube closure is generally found from mid-myelencephalon into the forebrain; often the walls of the diencephalon are apposed, although not fused; fusion of the telencephalic roof is variable. (3) The relative positions of such markers as the otocyst, optic cup, infundibulum, Rathke's pocket, and nasal pits indicate that cadmium treatment interferes with growth of the mesencephalon and diencephalon, normal establishment of the cranial flexure, and extension of the prosencephalon; positions of the sense organs relative to occipital and facial bones, as demonstrated in alizarin preparations of near-term fetuses, provide supporting evidence. (4) Formation of the eye rudiment frequently fails to proceed beyond the optic sulcus/early optic vesicle stage, presumably as a consequence of interference with normal growth processes in the diencephalon.

Plausible hypotheses to explain cadmium interference with neural tube closure may be developed on the basis of the importance of cell surface modifications and/or microfilaments and microtubules in the neurulation process,[28,32,33,50] although observations bearing directly on these possibilities have yet to be reported. Virtual restriction of cadmium interference with neural tube closure to the cephalic region, however, would require additional assumptions. This suggests the likelihood that cadmium is affecting some aspect of the closure process that is either peculiar to, or of overriding importance for, closure of the cephalic neural tube.

Growth processes in the cephalic neural tube are of particular interest as mediators of cadmium interference with neural tube closure. Intracellular damage and subsequent cell loss[51] may be of particular significance in high growth rate regions of the cephalic neuroepithelium. Abnormal cranial flexure, interference with optic vesicle formation, and malpositioning of various cephalic structures may all reflect interference with normal growth. Some caution is needed here in distinguishing between growth processes, interference with which may be responsible for failure of neural tube closure, and growth processes that are interfered with because the neural tube remains open. For example, elongation of the prosencephalon appears to depend on cerebrospinal fluid pressure since it does not take place in heads explanted into an in vitro culture system.[20] Failure of the neural tube to close then will contribute to malposition of markers within the prosencephalon, although it may not be solely responsible. Elongation of the mesencephalon and deepening of the cranial flexure are not dependent on neural tube closure, and may in fact contribute to the closure process.[33] Thus abnormal cranial flexure can be a consequence of cadmium interference with normal growth processes in the mesencephalon, and may in turn contribute to the failure of neural tube closure.

4. The Maternal Genotype and Neural Tube Malformations

Studies in the mouse, rat, and hamster have suggested a strong strain dependence of the developmental response to cadmium treatment.[1,17,31,39,53] Pregnant females from a random-bred colony of Swiss albino mice failed to litter following a single subcutaneous injection of cadmium (0.02 mmol/kg) on any gestation day from 6 through 17.[6] Actual pregnancy of animals treated on gestation day 11 or earlier was not confirmed in all cases, but microscopic sections of implantation sites taken from a small sample collected 24 h after treatment demonstrated the presence of some embryos showing various degrees of disintegration. Extensive studies with mice of the C57BL10/ChPr(B10) strain demonstrated substantial embryo mortality (80%) following injection of cadmium at 0.02 mmol/kg on day 11,[53] but less than 50% mortality was observed following treatment around the beginning of gestation day 8 or 9.[39] Mice of the NAW/Pr strain subjected to the same treatment regimes proved to be resistant to the embryolethal effects of cadmium, and crosses between the two strains indicated that sensitivity to cadmium-induced embryotoxicity showed a complex pattern of inheritance, including a maternal effect.[39,53]

Fetuses surviving cadmium treatment at the beginning of gestation day 11 showed no externally visible defects, but major craniofacial and eye defects were commonly seen among B10 fetuses surviving treatment at the beginning of gestation day 8.[37] Breeding experiments involving B10 and NAW/Pr demonstrated a major role for the maternal genotype in shaping the developmental response to cadmium, but the embryonic genotype can exert a modifying influence in specific situations.[39] When a phenotypic representation of the maternal contribution to cadmium-induced teratogenicity was constructed by assigning a teratogenicity score to each mother on the basis of the overall response of her litter to cadmium challenge, it became apparent that teratogenicity score was transmitted as a dominant trait and showed segregation in F_2 and BC_1 ($F_1 \times$ NAW/Pr) generations. That the mean teratogenicity score for BC_1 mothers differed depending on the sire provided evidence that the embryonic genotype also played a role in determining the response to cadmium. This role appeared to be relatively strong in the case of cadmium-induced interference with facial development, but less strong for interference with development of the neural tube. The role of the embryonic genotype in cadmium-induced interference with eye development was somewhat intermediate.

Examination of individual teratogenicity scores for BC_1 females, irrespective of sire, showed that 15 fell within three standard deviations (SD) of the mean score for NAW/Pr (12.3 ± 6.7) and 12 fell within three SD of the mean score for C57BL/10ChPr (69.3 ± 9.5).[39] Since data for BC_1 mothers represent an assay for segregation of genes responsible for the maternal influence, the apparent bimodal distribution could suggest segregation at a single locus. In an attempt to clarify the situation, a second backcross to NAW/Pr was made; female progeny (BC_2) were raised to maturity, bred to NAW/Pr sires, and their daughters (BC_3) tested as an indicator of the genotype of the BC_2 females (this progeny-testing approach was necessary because of the segregation in the BC generations). Data from 70 litters were obtained in order to classify 14 BC_2 females. Again, a bimodal distribution was suggested: teratogenicity scores for six females fell within three SD of the mean value for NAW/Pr; scores for all females fell outside of three SD for the mean score for C57BL/10ChPr. That the percentage of females classified as NAW/Pr-like in their response did not significantly increase from the $BC_1(48\%)$ to the $BC_2(43\%)$, a likely expectation if a single locus difference determined resistance/ susceptibility, makes it unlikely that a single locus hypothesis is correct. Although sample size was small in both tests, it seems likely that more than one genetic locus is involved.

The identity of the gene loci influencing the teratogenic response to cadmium has yet to be determined. At the present time any simple relationship between teratogenicity and *cdm*,[45,46] a locus defined on the basis of susceptibility/resistance to cadmium-induced testicular hemorrhagic necrosis[7,21] seems unlikely. Testicular cadmium resistance is recessive to cadmium susceptibility; in the case of the maternal influence on cadmium-induced teratogenicity, resistance is also recessive to susceptibility. However, data for cadmium-induced teratogenicity and embryo mortality are opposite to expectation on the basis of *cdm* genotype in B10 and NAW/Pr.[39,53] Data for cadmium-induced limb defects also demonstrate a lack of correspondence between *cdm* genotype and embryo susceptibility.[30,31]

The association between cadmium resistance of various cell culture lines and the synthesis of metallothioneins[8,12,18,22,25,41,44] suggests the possibility of roles for mouse metallothionein-I and/or metallothionein-II genes in the regulation of teratogenic susceptibility. Metallothionein-I *(Mt-I)* has been isolated and sequenced,[11,19] and appears to be a single copy gene in normal mouse cells. Recent studies have shown that murine Friend leukemia cells resistant to cadmium toxicity (Cd^r cells) have on a per cell basis six-fold more *Mt-I* genes than do nonresistant cells (Cd^s cells).[2] Chromosomal analysis indicates that Cd^r cells are mostly subtetraploid and contain 2–5 very small chromosomes not seen in Cd^s cells. Cd^s cells appear to be diploid. Although *Mt-I* amplification in the cadmium-resistant cells appears to be related to the karyotypic abnormality, this need not rule out the possibility of important variations at the *Mt-I* locus in inbred mouse strains. Correlations between induction of metallothionein and mortality from cadmium treatment reported for three inbred strains of mice [24,48] are consistent with this possibility. Studies of *Mt-I* in B10 and NAW/Pr mice have not yet been attempted.

Additional data (unpublished) bearing on the genetic control of cadmium-induced teratogenicity have been obtained by the use of inbred mouse strains maintained heterozygous for specific gene loci. These include NSP/Pr [segregation at the *Sp* (Splotch) locus], MSL/Pr [segregation of *Mi*^{wh} (Microphthalmic-white locus)]. Two approaches have been used: (a) reciprocal matings within an inbred strain (e.g., NSP/Pr: $Sp/\pm \times \pm/\pm$) and (b) outcrosses between inbred females and B10 males (e.g., $Sp/\pm \times$ B10; $\pm/\pm \times$ B10). Data obtained indicate effects of Sp on cadmium-induced teratogenicity, but Mi^{wh} appears to be without effect. The frequency of exencephaly in litters sired in $Sp/\pm$ mothers is greater than the frequency in $\pm/\pm$ mothers.

5. Cadmium–Zinc Interaction

Ferm and Carpenter[14,15] have reported that simultaneous injection of pregnant hamsters on the eighth day of gestation with equal amounts of cadmium and zinc resulted in a malformation rate of only 3% in comparison to 66% after injection with cadmium alone at 2 mg/kg body weight. A protective effect of zinc was also demonstrated if administered up to 6 h after the injection of cadmium, but not if 12 h elapsed between injections. Unpublished studies from this laboratory show that zinc also provides protection against the teratogenic effects of cadmium in the mouse. Simultaneous administration of zinc and cadmium at a molar ratio of two significantly reduced the frequency of total malformations in C57BL/10ChPr (B10) mice, but approximately one-half of the embryos still showed gross defects (Table 1). That facial defects were the most common abnormalities induced by cadmium in the

Table 1
Cadmium–Zinc Interaction[a]

	Treatment			
	Zn	Cd	Cd/Zn	Cd/Zn
Dosage, mmol/kg body weight	0.02	0.02	0.02/0.02	0.02/0.04
Litters examined	7	10	9	7
Implantation sites, total	66	83	77	61
Resorption sites, %	6	43	29	16
Abnormal embryos, %	3	89	62	53
Brain efects, total, %	0	68	40	33
With eye defects, %	0	68	33	23
Exencephaly, 5	0	47	26	11
Encephalocele, %	0	21	14	22
Eye defects, total, %	3	89	55	43
Isolated eye defects, %	3	21	22	20
Facial defects, total, %	0	30	13	10

[a]C57BL/10ChPr mice. Maternal ip injection on gestation day 8 at 8–10 AM; fetuses examined on gestation day 15.

study with hamsters may partially explain the apparent difference in effectiveness of zinc supplementation in the two experimental animals. In the mouse, zinc was most effective in protecting against cadmium-induced facial defects, but the frequency of facial defects in unsupplemented cadmium-treated litters was only 30% in comparison to frequencies of 68 and 89% for brain and eye defects, respectively. In the zinc-supplemented group the frequency of brain defects was reduced to 33%, and the ratio of encephalocele, a less severe brain defect, to exencephaly was greater in the group treated with both metals than in the group treated with cadmium alone (2.0 and 0.5, respectively). The frequency of eye defects was reduced to 43% in the group treated with both metals, but this was accomplished by a reduction only in the number of animals showing both brain and eye defects. The frequency of animals showing isolated eye defects was unaffected by zinc supplementation. The implications of these observations remain to be explored.

Further study of the zinc–cadmium interaction indicated that a more drastic reduction in the frequency of brain defects could be accomplished if supplementary zinc was injected at 8–11 h after the cadmium injection (Table 2). Under these conditions, the frequency of eye defects was reduced to the level characteristic for B10 mice. Comparison of data obtained when all of the zinc supplement was provided at 8 h post-cadmium injection with data

Table 2
Cadmium–Zinc Interaction: Timing Effects[a]

	Treatments			
	Cd[b]/Zn[c]	Cd/Zn[d]	Cd/Zn[d]	Cd/Zn[c]
Timing	8A/8A	8A/8A,3P	8A/8A,7P	8A/4P
Litters examined	7	12	10	7
Implantation sites, total	61	94	86	60
Resorption sites, %	10	16	20	15
Abnormal embryos, %	53	49	16	16
Brain defects, total, %	33	29	4	2
With eye defects, %	23	25	1	0
Eye defects, total, %	43	46	13	14
Isolated eye defects, %	20	20	12	14
Facial defects, total, %	10	6	0	4

[a]C57BL/10ChPr mice. Maternal ip injection on gestation day 8; fetuses examined on gestation day 15.
[b]Cd, 0.02 mmol/kg body weight.
[c]Zn, 0.04 mmol/kg body weight.
[d]Zn, 0.02 mmol/kg body weight, each injection, A = AM. P = PM..

Table 3
Cadmium–Zinc Interaction: Genetic Effects[a]

	Resorption sites, %	Brain defects, %	Eye defects, %	Facial defects, %
Embryonic genotype[b]: B10.B10				
Cd[c]	43	68	89	30
Cd/Zn-1[d]	29	40	55	13
Cd/Zn-2[e]	16	33	43	10
Embryonic genotype: B10.N (F_1)				
Cd	51	49	68	11
Cd/Zn-1	8	6	25	0
Cd/Zn-2	9	0	0	0
Embryonic genotype: B10.B10.N (RBC_1)				
Cd	53	38	57	18
Cd/Zn-1	19	36	55	12
Cd/Zn-2	17	13	28	0
Embryonic geyotype: B10N.B10 (RBC_1)				
Cd	30	70	66	26
Cd/Zn-1	20	48	58	19
Cd/Zn-2	3	10	10	1

[a]Maternal ip injection on gestation day 8 at 8–10 A.M.; fetuses examined on gestation day 15.
[b]Genotype of C57BL/10ChPr, B10; NAW/Pr, N. Maternal contribution on left by convention.
[c]Cd, 0.02 mM/kg bw.
[d]Zn, 0.02 mM/kg bw.
[e]Zn, 0.04 mM/kg bw.

obtained when one-half of the zinc was given with the cadmium and the other one-half was given either 7 or 11 h later suggested that a zinc-responsive period began around the middle of the 8th d of gestation. The extent of the zinc responsive period was not determined, but supplemental zinc given 24 h after cadmium injection early on the 8th d of gestation was ineffective. The nature of the zinc–cadmium interaction needs to be clarified, but these observations suggest that zinc becomes limiting in the cadmium-treated animals sometime during the 8th d of gestation. The teratogenicity of zinc deficiency has been well documented,[26,49] and is considered in detail in another chapter in this publication.[10]

Data presented in Table 3 suggest that the importance of the timing of supplemental treatment with zinc relative to the time of cadmium injection is dependent on the embryonic genotype. Differential responses to simultaneous injection of zinc and cadmium are elicited in inbred, $F_1(B10 \times NAW/Pr)$ and $RBC_1(B10 \times F_1)$ embryos developing in B10 mothers. F_1 embryos show no defects at all. Inbred embryos show the highest frequency of defects; BC_1 embryos, an intermediate level. The nature of the genotype–zinc–cadmium interaction is under study, but is unclear at the present time. RBC_1 embryos developing in F_1 mothers are less likely to show isolated eye defects than are RBC_1 embryos developing in inbred mothers, or RBC_2 embryos developing in RBC_1 mothers, but this is probably related to the inheritance of sporadic eye defects characteristic of C57BL/10ChPr.[4]

6. Summary

Cadmium has been shown to be teratogenic in hamsters, mice, and rats, under experimental situations involving either embryonic exposure *in utero* or challenge in an in vitro test system. Interference with normal development of the brain and eye rudiments is an important manifestation of cadmium teratogenicity, but malformations may also be observed in other soft tissues and in the skeleton. Specific defects observed are related to developmental stage at the time of exposure.

Investigations of cadmium teratogenicity in the central nervous system have for the most part focused attention on major defects that can be readily detected upon gross examination. Some histological observations have been reported, but information is still needed in order to bridge the gap between immediate cellular consequences of cadmium exposure and the resulting malformed embryo. More subtle aberrations, which may be of functional significance, are likely to be discovered by this approach. This would include, but not be limited to, behavioral consequences of cadmium exposure *in utero*. Behavioral studies have been reported, and attempts to relate these effects to cellular and tissue alterations in the developing brain initiated, but these are relatively few in number.

Teratogenicity studies with laboratory animals provide evidence of genetic control of the developmental response to cadmium. In the mouse, the maternal genotype plays a major role in determining responsiveness of the developing brain particularly. Identification of the gene loci involved has not been accomplished, but progress being made in cytological and molecular analysis of the mouse genome encourages increased attention to this task. Accumulation of information relative to mechanisms whereby specific cell systems exhibit resistance to cadmium upon exposure in culture suggests that an understanding of how the developing brain or embryo as a whole may come to terms with cadmium in its environment is a reasonable goal. Studies of the protective role of zinc in cadmium teratogenicity may contribute to this understanding. Dissection of the maternal and embryonic genotypes conditioning various aspects of the teratogenic response to cadmium may, however, provide the greatest impetus for progress in this area.

Acknowledgments

The following have participated in the research efforts reported from this laboratory: Jean S. Haines, John Miska (deceased), Timothy Charlesbois, and Katherine Maurer. Clerical and typing assistance were provided by Julie Harley, Elizabeth Pastor and Joan Landon.

Scientific Contribution No. 974 of the Storrs Agricultural Experiment Station.

References

1. Barr, M. Jr. (1973) The teratogenicity of cadmium chloride in two stocks of Wistar rats. *Teratology* 7, 237–242.
2. Beach, L.R. and Palmiter, R.D. (1981) Amplification of the metallothionein-I gene in cadmium-resistant mouse cells. *Proc. Natl. Acad. Sci. USA* 78, 2110–2114.
3. Butcher, R.E.(1976) Behavioral testing as a method for assessing risk. *Environ. Health Persp.* 18, 75–78.
4. Chase, H.B. (1942) Studies on an anophthalmic strain of mice. III. Results of crosses with other strains. *Genetics* 27, 481–503.
5. Chernoff, N. (1973) Teratogenic effects of cadmium in rats. *Teratology* 8, 29–32.
6. Chiquoine, A.D. (1965) Effect of cadmium chloride on the pregnant albino mouse. *J. Reprod. Fertil.* 10, 263–265.
7. Chiquoine, A.D. and Suntzeff, V. (1965) Sensitivity of mammals to cadmium necrosis of the testis. *J. Reprod. Fertil.* 10, 455–457.
8. Compere, S.J. and Palmiter, R.D. (1981) DNA methylation controls the inducibility of the mouse metallothionein-I gene in lymphoid cells. *Cell* 25, 233–240.
9. Cooper, G.P., Choudhury, H., Hastings, L. and Petering, H.G. (1978) Prenatal cadmium exposure: effects on essential trace metals and behavior in rats. In: D.D. Mahlum, M.R. Sikov, P.L. Hackett and F.D.

Andrew (Eds.), *Developmental Toxicology of Energy-Related Pollutants*, DOE Symposium Series 47, Technical Information Center, US Dept. of Energy, pp. 627–637.

10. Dreosti, I.E. (1982) Zinc and the central nervous system. In: I.E. Dreosti and R.M. Smith (Eds.), *Neurobiology of the Trace Elements*, Humana Press, Clifton, NJ.

11. Durnam, D.M. and Palmiter, R.D. (1981) Transcriptional regulation of the mouse metallothionein—I gene by heavy metals. *J. Biol. Chem.* **256**, 5712–5716.

12. Enger, M.D., Ferzoco, L.T., Tobey, R.A. and Hildebrand, C.E. (1981) Cadmium resistance correlated with cadmium uptake and thionein binding in CHO cell variants Cd^{r}20F4 and Cd^{r}30F9. *J. Toxicol. Environ. Health* **7**, 657–690.

13. Ferm, V.H. (1971) Developmental malformations induced by cadmium—A study of timed injections during embryogenesis. *Biol. Neonate* **19**, 101–107.

14. Ferm, V.H. and Carpenter, S.J. (1967) Teratogenic effect of cadmium and its inhibition by zinc. *Nature* **216** 1123.

15. Ferm, V.H. and Carpenter, S.J. (1968) The relationship of cadmium and zinc in experimimental mammalian teratogenesis. *Lab. Invest.* **18**, 429–432.

16. Gale, T.F. and Ferm, V.H. (1973) Skeletal malformations resulting from cadmium treatment in the hamster. *Biol. Neonate* **23**, 149–160.

17. Gale, T.F. and Layton, W.M. (1980) The susceptibility of inbred strains of hamsters to cadmium-induced embryotoxicity. *Teratology* **21**, 181–186.

18. Gick, G., McCarty, K.S. Jr. and McCarty, K.S. Sr. (1981) The role of metallothionein synthesis in cadmium- and zinc-resistant CHO-KIM cells. *Exp. Cell Res.* **132**, 23–30.

19. Glanville, N., Durnam, D.M. and Palmiter, R.D. (1981) Structure of mouse metallothionein-I gene and its mRNA. *Nature* **292**, 267–269.

20. Goodrum, G.R. and Jacobson, A.G. (1981) Cephalic flexure formation in the chick embryo. *J. Exp. Zool.* **216**, 399–406.

21. Gunn, S.A., Gould, T.C. and Anderson, W.A.D.(1965) Strain differences in susceptibility of mice and rats to cadmium-induced testicular damage. *J. Reprod. Fertil.* **10**, 273–275.

22. Hager, L.J. and Palmiter, R.D. (1981) Transcriptional regulation of mouse liver metallothionein-I gene by glucocorticoids. *Nature* **291**, 340–342.

23. Hastings, L, Choudhury, H., Petering, H.G. and Cooper, G.P. (1978) Behavioral and biochemical effects of low-level prenatal cadmium exposure in rats. *Bull. Environ. Contam. Toxicol.* **20**, 96–101.

24. Hata, A., Tsunoo, H., Nakajima, H., Shintaku, K. and Kimura, M. (1980) Acute cadmium intoxication in inbred mice: A study on strain differences. *Chem-Biol. Interactions* **32**, 29–39.

25. Hildebrand,m C.E., Enger, M.D. and Tobey, R.A. (1980) Comparative studies of zinc metabolism in cultured Chinese hamster cells with

differing metallothionein-induction capacities. *Biol. Trace Element Res.* **2**, 235–246.

26. Hurley, L.S. (1979) The role of zinc in prenatal and neonatal development. In: N. Kharasch (Ed.), *Trace Metals in Health and Disease*, Raven Press, New York, pp. 167–175.

27. Jacobson, A.G. and Gordon, R.M. (1976) Changes in the shape of the developing vertebrate nervous system analyzed experimentally, mathematically and by computer simulation. *J. Exp. Zool.* **197**, 191–246.

28. Karfunkel, P. (1974) The mechanisms of neural tube formation. *Int. Rev. Cytol.* **38**, 245–271.

29. Klein, N.W., Vogler, M.A., Chatot, C.L. and Pierro, L.J. (1980) The use of cultured embryos to evaluate the teratogenic activity of serum: cadmium and cyclophosphamide. *Teratology* **21**, 199–208.

30. Layton, W.M. Jr. and Layton, M.W. (1977) Genetic studies of cadmium-induced limb defects in mice. *Teratology* **15**, 22A.

31. Layton, W.M. Jr. and Layton, M.W. (1979) Cadmium induced limb defects in mice: strain associated differences in sensitivity. *Teratology* **19**, 229–236.

32. Messier, P.-E. and Seguin, C. (1978) The effects of high hydrostatic pressure on microfilaments and microtubules in *Xenopus laevis*. *J. Embryol. Exp. Morph.* **44**, 281–295.

33. Morriss, G.M. and New, D.A.T. (1979) Effect of oxygen concentration on morphogenesis of cranial neural folds and neural crest in cultured rat embryos. *J. Embryol. Exp. Morph.* **54**, 17–35.

34. Mulvihill, J.E.S., Gamm, S.H. and Ferm, V.H. (1970) Facial formation in normal and cadmium-treated golden hamsters. *J. Embryol. Exp. Morph.* **24**, 393–403.

35. Parizek, J. (1964) Vascular changes at sites of oestrogen biosynthesis produced by parenteral injection of cadmium salts: the destruction of placenta by cadmium salts. *J. Reprod. Fertil.* **7**, 263–265.

36. Parzyck, D.C., Shaw, S.M., Kessler, W.V., Vetter, R.J., Van Sickle, D.C. and Maytes, R.A. (1978) Fetal effects of cadmium in pregnant rats on normal and zinc deficient diets. *Bull. Environ. Contam. Toxicol.* **19**, 206–214.

37. Pierro, L.J. and Haines, J.S. (1976) Cd-induced exencephaly and eye defects in the mouse. *Teratology* **13**, 33A.

38. Pierro, L.J. and Haines, J.S. (1977) Maternal genotype and Cd-induced exencephaly and eye defects in the mouse. *Teratology* **15**, 23A.

39. Pierro, L.J. and Haines, J.S. (1978) Cadmium-induced teratogenicity and embryotoxicity in the mouse. In: D.D. Mahlum, M.R. Sikov, P.L. Hackett and F.D. Andrew (Eds.), *Developmental Toxicology of Energy-Related Pollutants*, DOE Symposium Series 47, Technical Information Center, US Dept. of Energy, pp. 614–626.

40. Rodier, P.M. (1976) Critical periods for behavioral anomalies in mice. *Environ. Health Persp.* **18**, 79–83.

41. Rugstad, H.E. and Norseth, T. (1975) Cadmium resistance and content of cadmium-binding protein in cultured human cells. *Nature* **257,** 136–137.

42. Rugstad, H.E. and Norseth, T. (1978) Cadmium resistance and content of cadmium-binding protein in two enzyme-deficient mutants of mouse fibroblasts (L-cells). *Biochem. Pharmacol.* **27,** 647–650.

43. Schneider, B.F. and Norton, S. (1979) Equivalent ages in rat, mouse and chick embryos. *Teratology* **19,** 273–278.

44. Takeuchi, Y.K., Sakai, H. and Takeuchi, I.K. (1979) Cadmium-induced congenital micro- or anophthalmia in rats. *Cong. Anom.* **19,** 113–123.

45. Taylor, B.A., (1976) Linkage of the cadmium resistance locus to loci on mouse chromosome 12. *J. Hered.* **67,** 389–390.

46. Taylor, B.A., Heiniger, H.J. and Meir, H. (1973) Genetic analysis of resistance to cadmium-induced testicular damage in mice. *Proc. Soc. Exp. Biol. Med.* **143,** 629–633.

47. Tsujii, H. and Hoshishima, K. (1979) The effect of the administration of trace amounts of metals to pregnant mice upon the behavior and learning of their offspring. *J. Fac. Agric. Shinshu U.* **16,** 13–27.

48. Tsunoo, H., Nakajima, H., Hata, A. and Kimura, M. (1979) Genetic influence on induction of metallothionein and mortality from cadmium intoxication. *Toxicology Lett.* **4,** 253–256.

49. Warkany, J. and Petering, H.G. (1972) Congenital malformations of the central nervous system in rats produced by maternal zinc deficiency. *Teratology* **5,** 319–334.

50. Waterman, R.E. (1976) Topographical changes along the neural fold associated with neurulation in the hamster and mouse. *Am. J. Anat.* **146,** 151–172.

51. Webster, W.S. and Messerle, K. (1980) Changes in the mouse neuroepithelium associated with cadmium-induced neural tube defects. *Teratology* **21,** 79–88.

52. Webster, W.S. and Valois, A.A. (1981) The toxic effects of cadmium on the neonatal mouse CNS. *J. Neuropath. Exp. Neurol.* **40,** 247–257.

53. Wolkowski, R.M. (1974) Differential cadmium-induced embryotoxicity in two inbred mouse strains. I. Analysis of inheritance of the response to cadmium and of the presence of cadmium in fetal and placental tissues. *Teratology* **10,** 243–262.

Chapter 4

Neurobiological And Behavioral Effects Of Lead

Ted L. Petit and Dennis P. Alfano

1. Introduction

The potential toxicity of many trace metals including aluminum, cadmium, mercury, lead (Pb), copper, zinc and arsenic is well-established. Among those trace metals posing a serious hazard to health, mercury, cadmium, and lead constitute the greatest risk from environmental exposure.[37]

Lead is an ubiquitous element with numerous industrial and domestic applications.[40] Considering the numerous sources of Pb that exist in our society, as well as the widespread increases in the Pb levels of our environment,[76] serious concern has recently been expressed over the adverse effects of Pb on human health (see Posner et al.[160]). For example, the National Bureau of Standards[138] estimates that over 600,000 children have blood levels equal to or greater than clinically acceptable levels. According to the US Environmental Protection Agency,[56] approximately 900,000 newborn children risk potential Pb intoxication via their mother. Occupational Pb exposure further increases the potential for Pb poisoning in adults. Estimates of the work force exposed to levels of Pb sufficiently high to result in clinical symptoms range from 10 to 35 percent.[137] It is evident that undue exposure to Pb poses a serious health hazard to infants, children, and adults. In this chapter we will focus on the neurobiological and behavioral effects of Pb.

2. Behavioral Effects Following Lead Exposure

2.1. Behavioral Effects In Adults

It is apparent from the literature that the numerous sources of Pb result in widespread environmental pollution and contamination of food and beverages.[40] Further, when exposure to Pb occurs in the workplace the potential for Pb intake increases considerably.[76] The prevalence of occupational Pb exposure in a wide range of American industries was well-established over 40 yr ago,[16] with estimates that approximately 1.5% of all workers were exposed to significant amounts of Pb. Similar estimates of occupational Pb exposure have been recently made in Finland[210] and Denmark.[75] Grandjean[76] has compiled a summary of those industrial settings that hold the potential for exposure to hazardous levels of Pb as well as those with less significant levels of exposure.

Not surprisingly, a large proportion of Pb-exposed workers suffer from symptoms characteristic of Pb intoxication.[118,119] Lane et al.[115] have listed the following as the major signs and symptoms of Pb toxicity in adults: tiredness, lassitude, constipation, abdominal discomfort or pain, anorexia, disturbed sleep, irritability, anemia, pallor, diarrhea, and nausea. Severe signs and symptoms can include severe abdominal pain, a reduction in muscle strength (e.g., wrist-drop), as well as other indicators of peripheral neuropathy and occasionally, encephalopathy. These major signs and symptoms of Pb poisoning have been summarized by Culver[44] based on the target organ involved, by Dagg et al.[45] who described these signs in descending order of frequency of the presenting symptom, and by the National Academy of Sciences[137] and the World Health Organization[226] in terms of the gross indicators of Pb toxicity and toxicokinetics of Pb absorption.

Measurements of blood Pb levels have been traditionally regarded as the best indicator of the risk of Pb poisoning[226] and the most meaningful in monitoring workers exposed to Pb.[93] Manifest symptoms of Pb toxicity in the occupationally exposed worker rarely occur at blood Pb levels below 80 μg/dL.[35,226] In a survey of 26 industrial operations conducted by the National Academy of Sciences,[137] 10–35% of occupationally exposed workers had blood

Pb levels of 100 μg/dL or more. Evidence from other sources (see Grandjean[76]) has suggested that measurement of zinc protoporphyrin (ZPP) and, at lower exposure levels, delta-aminolevulinic acid dehydratase (ALAD) may have higher predictive validity than blood Pb analysis. Additional biological indices of Pb exposure have also been discussed by Posner et al.[160]

Standardized psychological and neuropsychological tests have also been used to assess the behavior and cognitive functioning of workers occupationally exposed to Pb. Repko and Corum[166] have previously reviewed the behavioral testing methods and results obtained in these investigations. Among the more recent studies, Grandjean et al.[78] observed deficient long-term memory, as well as problems in verbal and visuo-spatial abstraction and in psychomotor speed in a group of occupationally exposed workers with blood Pb levels ranging from 12 to 88 μg/dL. In this investigation, both blood Pb and ZPP correlated significantly with performance on most tests in both the exposed group and in the total population studied. Arnvig et al.[9] also observed deficient memory and psychomotor performance, as well as poor attention and concentration abilities in workers with blood Pb levels ranging from 58 to 82 μg/dL. Valciukas[213,214] found the deficiencies in visuo-spatial and visuo-perceptual abilities in their Pb-exposed group to be correlated with ZPP to a greater degree than with blood Pb level.

Personality changes have also been observed in adults following exposure to Pb. An increased prevalence of depression, hostility, and general dysphoria has been observed by Repko et al,[168] while Hanninen et al.[83] found that low level exposure to Pb may induce disturbances in subjective feelings of well being.

2.2. Behavioral Effects in Infants and Children

The signs and symptoms of Pb poisoning in infants and children are somewhat different from those usually found in adults. For example, though peripheral neuropathy is more frequently observed in adults, encephalopathy is more commonly found in children.[73] The onset of Pb poisoning in children can be insidious, with the symptoms nondescript: irritability and anorexia may be followed by vomiting, abdominal pain, and ataxia, with headache, drowsiness, speech problems, seizures, fever, dehydration, stupor, and coma occurring in the more severe stages of encephalopathy.[141] Sachs et al.[172] have described the signs and symptoms of Pb poisoning in children based on the descending order of frequency as being: drowsiness, irritability, vomiting, gastrointestinal symptoms, ataxia, stupor, and fatigue.

Since permanent damage to the CNS may occur, acute encephalopathy is one of the most serious consequences of Pb poisoning. The American Academy of Pediatrics[7] estimates that of those children experiencing Pb encephalopathy, 25% will suffer permanent damage to the CNS. Fortunately, awareness of the problem has reduced the incidence of Pb encephalopathy through legislation restricting the allowable levels of Pb in paint (historically a major source of Pb for ingestion by children) and through routine blood Pb screening programs.[35] Standards suggested by the Center for Disease Control[35] associate undue Pb absorption in children with blood Pb levels of 30–69 μg/dL in the absence of clinical symptoms. Lead poisoning is defined as a blood Pb level equal to or greater than 70 μg/dL. As Needleman[141] points out, 1 g of paint containing 1% Pb contains 10 mg Pb or over 10 times the maximum permissible daily adult intake of Pb. Such small amounts taken over periods even as short as 3 months can produce clinical signs of Pb poisoning. The probability of such ingestion is high in children with pica (eating nonfood items), a common behavior in young children.

An early longitudinal investigation by Byers and Lord[29] reported the neurological and psychological sequelae of acute lead encephalopathy in 20 children. Many of the children showed short attention spans, had difficulty in spatial tasks and were impulsive, and all but one were failing in school. Numerous instances of recurrent seizures, mental retardation, and permanent brain damage following exposure to severely high levels of Pb have subsequently been described.[1,39,81,121,149,152,170] Perlstein and Attala[152] reported that 39% of their population of 425 cases of Pb-poisoned children displayed neuropsychological sequelae. Of those children with encephalopathy, 82% had sequelae, most notably mental retardation and epilepsy. Thus, high levels of Pb exposure can cause a number of moderate to severe long-term neurobehavioral deficits.

Several studies have suggested that chronic exposure to lower levels of Pb may also be associated with learning disabilities, problems in spatial integration and motor coordination, and various maladaptive behaviors including hyperactivity and an increased impulsiveness, distractibility, and aggressiveness.[48,114,122,139,142,15] The results of a number of these investigations[49,84] have led to the speculation that Pb may be an important etiological factor in some cases of hyperkinesis. Hyperkinesis in children is a complex syndrome characterized by a high level of motor activity, short attention span, low frustration tolerance, hyperexcitability and poor impulse control.[49] As such, widespread concern is currently being raised over the possible effects of chronic exposure to low levels of environmental Pb on infant and child behavioral development.[122,123,139,141]

Psychological assessments of children exposed to moderate levels of Pb have been carried out by a number of investigators. De la Burde and Choate[51] found a 3–5 point drop in the IQ scores of children with blood Pb levels above 40 μg/dL. They also observed an increase in mental retardation and behavioral difficulties, particularly short attention span. Similar results suggesting a drop in IQ of a few points with moderate levels of Pb exposure (40–80 μg/dL) were reported by other researchers.[104,114,140,151,170] Rummo et al.[170] found the IQ to be inversely proportional to the blood Pb level; children with levels of 40–60, 60–80, and 80 μg/dL having mean IQs of 92, 89, and 50, respectively. Deficits in motor coordination, visuo-motor coordination, reaction time, and attention and concentration are also seen in these children.[1,11,140] The research area of Pb exposure in children is quite extensive; the reader is referred to more detailed reviews specifically in this area.[19,112,141,171] In general, research to date indicates that exposure to encephalopathic levels of Pb results in a high incidence of severe neurobehavioral deficits. Exposure to moderate levels of Pb (blood levels of 40–80 μg/dL) are associated with a decline in IQ of several points and may cause some behavioral alterations, e.g., alterations in fine motor and visuo-motor coordination, although the exact nature and likelihood of occurrence of such alterations is less certain. Bornschein et al.[19] reviewed 16 studies looking for behavioral effects of moderate Pb exposure. Six studies he reviewed reported such changes, eight failed to find any effects, and two reported findings that were ambiguous with respect to behavioral changes. The psychological consequences of exposure levels of Pb leading to blood levels less than 40 μg/dL are even less well-established.

2.3. Behavioral Effects In Laboratory Animals

Deficient learning abilities,[23,25,198] deficits in the ability to solve complex spatial mazes[4] and impairments in visual discriminations,[53,62] motor coordination,[146] and reflex development have been reported in postnatally Pb-exposed animals,[147] all of which may be analogous to the lowered IQ and deficits in motor and visuo-motor coordination seen in Pb-exposed children. Similar to increased incidence of paroxysmal disorders in humans, recurrent seizures,[6] increased sensitivity to the behavioral effects of certain convulsant agents,[194] and alternations in maximum electroshock seizures[63] have been observed in experimental animals. Concurring with increased impulsiveness and emotionality in Pb-exposed humans, an increase in aggressiveness has been observed in laboratory animals. Hyperactivity has been observed following postnatal exposure to Pb,[53,54,71,153,176,192,217] although instances of both hypoactivity[79,85] and unaltered activity levels have been reported.[105] Increased activity appears to be dose- and situation-specific (unpublished observations) and may correspond to the hyperkinesis some researchers have reported in the human.

A series of behaviors suggesting deficient response inhibition is also seen following developmental Pb exposure. Specific behavioral deficits are seen on tests of passive avoidance,[153] one way active avoidance,[146] differential reinforcement of low rates of responding (DRL-20)[4] and fixed interval schedules of operant controlled behavior[228]; operant extinction and operant response inhibition,[146] spontaneous alternation,[95,100] and various discrimination reversal problems[28,53,146,169,199] have also been observed following developmental Pb exposure.

The above studies suggest that exposure of the developing laboratory animal to Pb can result in both subtle and specific behavioral indicators of CNS dysfunction. Unfortunately, a lack of standardization of the time, duration, method, and level of Pb exposure, the species and strain used and the method of behavioral assessment, as well as the presence of undernutrition at high levels of Pb exposure have made the development of unified formulation of Pb-induced neurobehavioral deficiency difficult. The interested reader is referred to Bornschein et al.[20] and Jason and Kellogg[94] for more extensive reviews of the behavioral effects of Pb exposure in laboratory animals.

Several behavioral responses of Pb appear to conform to a dose–response relationship. For example, Petit and his coworkers[4,153] have shown that the behavioral alterations of animals exposed to moderate levels of Pb are generally intermediate between those seen in high Pb-exposed animals and controls. Therefore, the differences in levels of Pb exposure may also explain some of the discrepant behavioral findings (e.g., activity changes) observed in different studies.

A number of behavioral changes observed following Pb exposure are very similar to those seen following hippocampal lesions.

Petit and his coworkers[4,154] have recently outlined a number of behavioral and anatomical alterations that suggest that the hippocampus may be the major site of action for Pb's neurobehavioral toxicological actions. Although it is clear that Pb alters other brain areas as well, a number of points to be discussed below support such a hypothesis.

Another major set of behavioral changes similar to those seen in the Pb-exposed animal is that seen after interruption of the cholinergic system. Disruption of either the hippocampal formation or the cholinergic system yields behavioral effects similar to those seen following developmental Pb exposure (see Alfano and Petit[4] for a review). A disruption of the cholinergic system is also consistent with neurochemical and electrophysiological studies to be discussed below.

Thus, the numerous behavioral changes observed in experimental animals exposed to Pb appear to correlate well with the behavioral disinhibition frequently seen after hippocampal–cholinergic disruption, and are consistent with anatomical, electrophysiological, and neurochemical research findings.

3. Mechanisms of Lead Entry and Storage

3.1. Peripheral Absorption of Lead

The rate of Pb absorption by the gut varies greatly with age. The developing animal (rat, mouse, and human) absorbs from 50 to 70% of ingested Pb.[2,18,102,216] A developmental decrease in absorption occurs at 20–22 d in the rat,[61] such that absorption in the adult rat is less than 1%.[18] Absorption in young adult humans has been reported to range from 5 to 14%.[92,97,163] The increased Pb absorption in neonates is further augmented by immature renal function, resulting in low excretion rates, and yielding animals with significantly higher levels of circulating Pb than adult animals after the same exposure.

Lead appears to compete with other divalent ions for absorption, as well as other physiological mechanisms. Of these ions, calcium (Ca), and to a lesser extent zinc (Zn) and other divalent ions, have received the most attention; their interaction has recently been reviewed by Mahaffey and Michaelson.[126]

Deficient Ca in the diet causes both increased absorption of Pb as well as increased toxicological effects of Pb. Researchers[13,127] have reported a fourfold increase in blood Pb content in Pb-exposed animals consuming low Ca diets. Other researchers[127,130,162,211,212] have shown that the amount of Pb absorbed depends on the level of Ca in the diet. Compared to the normal range, low levels of Ca cause an increase in the amount of Pb absorbed, while high levels of Ca cause a decrease in the amount of Pb absorbed. Further, these alterations in Ca levels are associated with changes in the severity of clinical or pathological signs of Pb toxicity.

The influence of Ca appears to be mediated at the level of gastrointestinal absorption. Low levels of Ca affect the toxicity of ingested, but not injected, Pb,[162] and Pb absorption decreases in ligated intestinal loops with increasing Ca concentration.[13,15] Thus, it appears that Pb and Ca compete for binding sites on absorptive proteins in the intestine.

Results similar to those with Ca have been found with Zn. A number of studies have shown that increased levels of Zn decrease the toxicity or blood level of Pb.[36,205,222] Research by Cerklewski and Forbes[36] indicated that injected Zn did not protect against Pb toxicity, suggesting Zn, like Ca, competes with Pb for absorption from the intestinal tract. Further research showed that injected Zn was capable of blocking the Pb effect on urinary aminolevulinic acid excretion, suggesting that Pb and Zn also compete at a physiological level beyond the gastrointestinal tract.[205] Related to this, Pb has been shown to depress Zn concentrations in the maternal milk of Pb-exposed dams.[18] Perhaps most relevant is that an increase in brain Pb concentration is associated with a concomitant decrease in brain Zn concentration.[144]

Lead has been postulated to be retained in three body compartments. In the blood, where about 95% of the Pb is bound to erythrocytes, Pb has a biological half-life of about 25–30 d.[163] In soft tissue, the second compartment, Pb has a half-life of a few months, although the half-life in the brain may be somewhat longer.[78,163] In the third compartment, calcified tissues, where more than 90% of body Pb is accumulated, the half-life in the tissues is 30–40 yr.[14,177] Several studies have shown that brain Pb remains unchanged as blood Pb levels drop, underscoring that circulating blood Pb levels are not likely to be a reliable index of Pb stored in the brain,[70,165] but rather indicate circulating levels of Pb at the time of sampling.

3.2. Central Nervous System Absorption of Lead

Massive entry of Pb from the blood supply into the brain is initially inhibited by the blood–brain barrier (BBB) or blood–nerve barrier (BNB). Stumpf et al.[201] have shown that injected ^{210}Pb is initially concentrated in blood vessels and lining cells of the brain, such as vascular endothelial cells, glial cells, choroid epithelium, meningeal cells, and ependymal cells in some regions. The authors suggest that Pb is taken up and retained by cells that are known to be components of the BBB. Lead does accumulate in the brain and peripheral nerves over time, however, with the accumulation of Pb in the brain being proportional to its concentration in the blood.[70] This accumulation appears to be progressive and more stable than underlying fluctuations in blood Pb concentrations, further pointing out that blood Pb concentrations serve as an indication of circulating Pb only at the time of sampling.

Research has suggested that developing animals are more susceptible to brain accumulations of Pb than adults.[86] For example, the amount of Pb accumulated in the brain is three times higher after the same Pb exposure in 10-d-old monkeys than in adult monkeys, and in rat pups than in their mothers eating the same diet (see Hertz et al.[86] for review). Trypan blue injected into the circulatory system of the encephalopathic Pb-exposed rat pups leaks into the brain, suggesting increased vascular permeability in Pb encephalopathy.[150] This breakdown in the BBB caused by Pb encephalopathy would then allow further amounts of Pb to pass more freely into the brain in developing animals, so playing an important role in the development of Pb-induced encephalopathy.[70]

Hertz et al.[86] have recently concluded that the BBB in the adult, in contrast to the newborn, is not as appreciably affected by Pb exposure. The BBB in the infant breaks down by 18–25 d, heralding an encephalopathy. Only after 70 d of Pb exposure in the adult did Windebank and Dyck[224] find a 15-fold increase in the rate of ^{210}Pb accumulation in the endoneurium. Segmental demyelination, however, began to appear after only 35 d of exposure. Therefore, the authors suggest that a simple hypothesis of a sudden breakdown of the BNB to Pb in the adult, allowing the entry of Pb-rich edema fluid, is not adequate to explain the accumulation of Pb in the adult neural tissue. Rather, Pb appears to cross into neural tissue immediately and to accumulate, causing neurotoxic effects. After a considerable period of accumulation the BNB or BBB breaks down allowing a further, dramatic influx of Pb. Thus, there may be a dose- and equilibrium–dependent saturation point of the barrier beyond which Pb penetrates into nervous tissue more rapidly.

The increased BBB permeability in newborns compared to adults, in addition to differences in absorption rate from the gut and excretion from the kidney may help explain why developing animals are more susceptible to Pb exposure.

Lead is not evenly distributed throughout the brain.[77] The largest accumulation of Pb in the normal animal is in the hippocampus, where its concentration is seven times greater than its concentration in the whole brain, or ten times greater than its concentration in the rest of the brain (without the hippocampus).[60] Further, the hippocampus, which comprises 13% of the fresh brain weight, contains about 50% of the total Pb, indicating a clear preferential accumulation of Pb by the hippocampus. The same laboratory later showed that Pb is present in high concentrations in the hilus of the dentate gyrus (about three times the average concentration for the hippocampus as a whole) and is also likely stored in high concentrations in the mossy fiber pathway, forming a partial basis, along with other metals, for the Timm's silver sulfide staining of these regions.[46]

In Pb-exposed animals, Klein and Loch[98] showed that Pb is taken up into the cerebellum to reach greater concentrations than in the cerebral cortex or brainstem (they did not examine the hippocampus separately). Dubas and Hrdina[55] found regional Pb concentrations to be highest in the hypothalamus, followed by the striatum, cortex, and midbrain in Pb-exposed animals; they also did not examine the hippocampus or cerebellum. Hrdina et al.[90] reported that, at 3 weeks, the Pb content of Pb exposed rats was highest in the cerebellum followed by the pons-medulla, hypothalamus, hippocampus, striatum, cortex, and midbrain. After 8 weeks of continued exposure the Pb content in the cerebellum, pons-medulla, and hypothalamus had dropped, while Pb in the other brain regions had increased. The highest concentrations at that time were in the pons-medulla and hippocampus. Though Modak et al.[133] did not find any significant differences in their brain regional Pb analysis in Pb-exposed rats, an examination of their data reveals that the hippocampus had the highest mean Pb concentration across the three Pb treatment groups. Thus, it appears that Pb accumulates predominantly in the hippocampus, but with the breakdown of the BBB with high levels of Pb, Pb-rich edemic fluid enters, giving other brain areas, particularly the cerebellum, very high concentrations of Pb. With time cerebellar Pb decreases so that the hippocampus once again has the highest concentration of Pb.

Using a single tracer dose, Pb accumulates largely in blood vessels and their lining cells.[90] As would be expected from a single dose, little Pb penetrated the BBB into neurons, but that which did penetrate was concentrated in the nucleus. Krigman et al.[109] reported that Pb accumulated in the following subcellular brain fractions in developing high Pb-exposed rats: 16.5% in mitochondrial supernatant, 34.4% in mitochondria, 30.8% in the nuclear fraction, 3.4% in synaptosomes, and 2.4% in myelin.

4. Neurochemistry of Lead Exposure

A great deal of research has been generated on the effects of Pb on neurochemical systems, with most work focused on the neurotransmitter systems for acetylcholine (ACh), norepinephrine (NE), dopamine (DA), serotonin (5HT), and gamma-aminobutyric acid (GABA). Several reviews of the neurochemical effects of Pb have recently been published.[68,69,90,183,1]9[3]

4.1. Cholinergic System

Studies on the neurochemical effects of Pb on the cholinergic system have assessed the effects in terms of (1) steady-state levels of ACh and its precursor choline (Ch), (2) cholinergic enzymes, (3) rate of ACh release and turnover, (4) synaptosomal transport mechanisms, and (5) pharmacological responses of Pb-exposed animals.

Steady-state levels of ACh and Ch have been examined by many investigators. Steady-state levels of ACh have been shown to be unchanged,[32,133,181,182,192] or increased,[91,133] while steady-state levels of choline have been shown to be either unchanged[32,181,182] or decreased.[181,182] Hrdina et al.[90] point out that the differences found by different investigators in the level of these chemicals likely stem from methodological differences, particularly the method of sacrifice, since there are rapid postmortem alterations in the levels of these substances.

The activity of choline acetyltransferase, which is responsible for the synthesis of ACh from Ch has been observed to undergo no change[32,133,134] or to show a 14–18% increase[133] in various brain areas. Acetylcholinesterase (which hydrolyses ACh and converts it back into Ch) activity, however, shows either no change,[32,91,133,199] or a decrease in activity[133,199] depending on brain areas and species examined. Again, these differences may be attributable to methodological differences between laboratories.

The most consistent effects of Pb on neurochemical systems have been found in ACh release and turnover rates. The results of several studies clearly point out that Pb exerts an inhibitory influence on ACh release from mouse cortical minces.[32] Lead also causes a strong decrease in ACh turnover rate with reported values ranging from 33 to 54% in different brain areas, with the greatest decrease occurring in the hippocampus.[181,183] Lead exposure in vivo and in vitro is also known to inhibit nerve terminal transport of choline.[17,186,192]

Both methylphenidate and amphetamine produce paradoxical or activity-attenuating responses in Pb-treated animals.[192,199] Although these drugs are generally considered catecholaminergic stimulants, amphetamine has been shown to increase ACh release. Further, methylphenidate was shown to reverse completely the Pb-mediated decrease of ACh,[32] and Shih et al.[184,185] have also shown a cholinergic effect of methylphenidate.

Thus, taken together, the neurochemical evidence strongly indicates that Pb causes a marked inhibition of cholinergic functioning, particularly the stimulated release of ACh from the presynaptic terminal, an effect that may be greatest in the hippocampus. Further, the therapeutic effects of amphetamine and methylphenidate on Pb-exposed animals may be mediated through a reversal of this Pb-induced ACh inhibition.

4.2. Biogenic Amines

The effects of Pb on steady-state levels of the catecholamines have been examined with differing results by different investigators. Steady states of DA may be somewhat decreased, although most researchers have found no changes in DA levels (see Shih and Hanin[182] and Silbergeld and Hruska[193] for reviews). There is also uncertainty about NE levels, although most research suggests either no change or an increase in NE levels.[182,183,193] Catecholamine synthesis and turnover rates have been reported to be either unaltered or to show an increase.[131,193] There have been consistent reports, however, of large elevations in the two catecholamine metabolites, homovanillic acid (HVA) and vanillylmandelic acid (MVA) in chronically Pb-exposed animals.[187] Further, dopamine reuptake by synaptosomes is inhibited in Pb-exposed rats and mice,[192,223] which would serve to increase dopaminergic neurotransmission. Also Pb is able to increase synaptosomal release of DA.[17,186] These results taken together suggest an increase in presynaptic catecholaminergic function in Pb-exposed animals.

Unlike the catecholamines, little evidence exists for any effect of Pb on 5HT steady-state levels, on high affinity uptake, or on levels of its metabolite 5-hydroxyindolacetic acid.[182]

4.3. GABA

Levels of GABA do not appear to be effected by Pb, with the possible exception of the cerebellum.[158,194,200] Silbergeld et al.[194] found a decreased high affinity uptake of GABA by synaptosomes, and decreased resting and stimulated release of GABA. Silbergeld and Hruska[193] further report that chronic Pb-exposure is associated with increased postsynaptic receptor binding of ^{3}H-GABA, suggesting a denervation-induced increase in receptor activity. This, in combination with a failure to induce such changes in vitro, suggested to them that the effects of Pb on the GABA system are secondary to the actions on other transmitter systems, such as ACh or the catecholamines.

5. Electrophysiological Effects

5.1. Electroencephalographic Changes

Electroencephalographic (EEG) recordings in children exposed to high levels of Pb have shown a number of nonspecific abnormalities, which appear to be related to the level of Pb exposure. Abnormalities such as slow persisting dysrhythmic activity, slowed alpha frequency, reduced amplitudes, and focal paroxysmal discharges have been reported.[22,27,30,57,196,208] Smith et al.[196] observed no significant alterations in the EEGs of children exposed to low levels of Pb (below 60 μg/dL).

In adult humans exposed to elevated levels of Pb, EEG recordings have shown a slowing of the alpha rhythm and increases in the amount of slow wave activity in the theta and delta range, diffuse irregular patterns, and focal paroxysmal discharges.[136,195] Saito[174] reported increased activity in theta and decreased activity at higher frequences. These EEG abnormalities can be reversed by chelation therapy.[195]

Adult animal studies have also shown abnormal EEG recordings following Pb exposure, with the type of abnormality varying with the degree of exposure. Recordings on tetraethyl lead (TEL)-exposed rabbits were characterized by reduced frequency and amplitude after relatively low Pb exposure, and diffuse dysrhythmic activity and paroxysmic activity after higher and more recent Pb exposure. Lead poisoning in dogs is associated with irregular, high voltage activity in the delta range,[229] while rats exposed to TEL show increased theta and to a lesser extent, alpha wave activity,[174] and changes in the duration and stability of rapid eye movement (REM) sleep.[227]

5.2. Evoked or Event-Related Potentials

Visual evoked potentials (VER) in rats exposed to low levels of Pb during development were examined by Fox et al.[62-64] An increase in the latency and a delay in the emergence of the primary (P1 N1) and secondary (P2 N2) components of the VER were observed during development. Significantly longer latencies were also observed in these animals tested as adults. Adult animals also showed an increased latency of the P1 wave in response to the second of a pair of flashes and a decreased ability to follow repetitive light flashes compared to controls, which the authors interpret as decreased CNS recoverability. Slow-wave voltage during sensory conditioning has been observed to vary as a linear function of blood Pb level.[145]

5.3. Maximal Nerve Conduction Velocity

With elevated blood Pb, either maximal nerve conduction velocity (MCV), or MCV of some nerve fibers, has been repeatedly shown to be decreased in adult humans when examined in the ulnar and radial nerves,[180] lateral popliteal nerve,[33] peroneal nerve,[59] median peroneal and sural nerve,[26] median and ulnar nerves,[179] peroneal, posterior tibial, median, and ulnar nerves,[167] median and posterior tibial nerves,[8] peroneal, median, and sural nerves,[26] and medial, ulnar, and posterior tibial nerves.[178] Slowed MCV has been observed in individuals without overt signs of Pb neuropathology, and blood Pb levels below 70 μg/dL with some individuals showing impairment at blood Pb levels of 40–50 μg/dL.[8,167,178,179] Although some researchers have found the maximum conduction velocity related to the maximum or average level of Pb in the blood,[8,178] others have not found the effects of Pb to follow a dose–response curve.[26]

MCV has also been found to be reduced in children exposed to elevated levels of Pb,[58,59,113] although this relationship has been challenged by some (see Posner et al.[160] for discussion).

5.4. Synaptic Transmission

In 1957, Kostial and Vouk[103] demonstrated that Pb blocked synaptic transmission in the superior cervical ganglion of the cat, and produced a decreased release of ACh during stimulation of the preganglionic fiber. Lead did not change the sensitivity of the ganglionic cells to injected ACh, further confirming that the effect was presynaptically mediated. Since that time, several investigators have shown that Pb reduces or blocks the postsynaptic response to presynaptic nerve stimulation when examined in frog, rat, and mouse neuromuscular junction.[21,99,128,188,189] This research has further shown that Pb does not alter the postsynaptic response to applied acetylcholine, indicating that the results cannot be explained by postsynaptic inhibition. All of these results are consistent with the interpretation that Pb reduces or blocks the stimulated release of ACh from presynaptic terminals.

Lead is also known to increase miniature end-plate potential frequency at the unstimulated neuromuscular junction, indicating that Pb causes an increase in spontaneous transmitter release.[128] These results, in contrast to the blocking effect of Pb on transmitter-stimulated release, suggest that there may be two mechanisms, one for stimulated and one for spontaneous transmitter release.

Little research to date has centered on the catecholamine synapses. Cooper and Steinberg[43] showed that Pb reduced the response to sympathetic nerve stimulation, but did not effect responses to applied norepinephrine or to direct muscle stimulation. These results suggest that Pb reduces the release of norepinephrine from the presynaptic adrenergic terminal, similar to its effects at the cholinergic terminal. Arguments put forth by Taylor et al.[202] however, suggest that Pb exerts a postsynaptic inhibitory effect on cerebellar NE synapses at levels far below those required to block ACh release. They propose that Pb exerts its effects on NE neurons postsynaptically by inhibiting adenylate cyclase. If Pb does block postsynaptic NE function, then the increased NE turnover rates and increased concentrations of catecholamine metabolites may reflect a compensatory mechanism. Thus, at present it is difficult to state with certainty the effects of Pb on the adrenergic system. The majority of the neurochemical research, however, supports a conclusion of increased presynaptic function of this system and this is consistent with the hypothesis of Taylor et al.[202] concerning different effects of Pb on cholinergic and adrenergic systems (see Silbergeld and Hruska[193]).

5.5. Competition with Calcium

The competition of Pb with Ca outside the CNS has already been discussed; research has shown that Pb strongly competes with Ca at the neuronal level as well. In the cholinergic system, the effects of Pb (e.g., the blocking of stimulated ACh release) are similar to those of lowered Ca.[32,100,103,186,189] Also, the effects of Pb and lowered Ca are additive when examined in vitro[186,189] and the effects of Pb can be partially or completely reversed by increasing Ca levels.[100,103] These findings, taken together, suggest that Pb exerts its effects on the cholinergic system by competing with Ca at the neuronal membrane level. Since Ca is necessary for the normal release of transmitter substances from the presynaptic nerve terminal, the effect of Pb would be to disrupt Ca-dependent transmitter release.

The effect of Pb on the dopaminergic system is less clear, and may be more complex. The effects of Pb on the dopamine system are not similar to those of calcium deficiency, and are not reversed by increased Ca (see Silbergeld and Hruska[193]). Silbergeld and her colleagues[193] postulated that Pb exerts differing effects at the dopaminergic and cholinergic nerve terminals because catecholamines, unlike ACh, are able to chelate Pb, thus allowing it to enter the nerve terminal. There, Pb may interact with binding sites in mitochondria. According to their theory, Pb would then exert its effect at dopaminergic terminals by binding internal cellular Ca, causing more external calcium entry during depolarization and resulting in turn in increased dopamine release. Clearly, further research is needed to confirm their interesting hypothesis and fully understand the mechanisms of Pb–Ca action at the catecholamine synapse. Nonetheless, their hypothesis still involves competition between Pb and Ca (albeit internal rather than external calcium), which is consistent with other research.

5.6. Subcellular Mechanisms: Integration

The data gathered to date suggests that Pb causes a clear inhibition of cholinergic functioning. This is observed at both the neurochemical and electrophysiological level. The effect appears to be mediated through competition with calcium, blocking the presynaptic-stimulated release of ACh. The effect on the catecholamines appears to be less well understood; although the results are not as clear, there appears to be a general trend of elevated activity of catecholaminergic functioning. This may be mediated through an internal binding of mitochondrial Ca, producing greater transmitter release, or the effect may be secondary (compensatory) to a disruption of postsynaptic functioning caused by blocking of adenylate cyclase.

It has been suggested that Pb causes a decrease in cholinergic functioning, which causes a corresponding increase in catecholaminergic functioning.[68,185,192] This hypothesis assumes a balance between cholinergic and aminergic systems.[12] Lead is thought to cause an inhibition of the cholinergic system, which then disinhibits the aminergic system causing activity to increase above normal levels. The pharmacological evidence gathered with amphetamine and methylphenidate support this hypothesis. Thus, the effects of Pb on the cholinergic system appear to be direct, while its effects on the aminergic system may be both direct and indirect.

6. Neuroanatomical Effects of Lead

6.1. General or Gross Changes

A number of neuroanatomical changes are associated with exposure to high levels of Pb during development, particularly levels that produce overt encephalopathy. Most of these changes are either not observed or have not been examined in the adult. The most frequently observed changes are an increase in brain edema causing a general swelling of the brain that has been observed in several species including rodents and primates and is generally considered most obvious in the cerebellum.[41,89,132,149,150,161] Hemorrhagic lesions, particularly in the cerebellum, are also generally seen following high Pb exposure.[89,107]

Further analysis shows reduced brain size or weight ranging from 9 to 18% following Pb exposure in the encephalopathic range to no differences with low levels of Pb.[5,106,107,108,153,155] Reduced weights of various brain areas including the cerebral cortex, cerebellum, and hippocampus have also been reported.[110,132,153] Of those areas examined, the cerebellum and hippocampus appear to suffer the greatest loss.[132,153] Petit and Alfano[153] reported a weight analysis of various brain regions that revealed that the hippocampus suffers the greatest percent weight loss of all brain structures examined, including the cerebellum. The reduction in brain weight is found only after exposure to high levels of Pb that also retard body growth. Reductions in brain weight are not found either with lower levels of Pb or even with high Pb levels that produce an encephalopathy if steps are taken to insure adequate nutritional status for normal weight gain[3] or if body weight is treated as a covariate.[5]

Histological analysis reveals a proliferation of astroglial cells in animals exposed to high Pb during development.[107] Histological studies also confirm brain weight analysis, in that there is a reduction in the thickness of several brain areas. The cerebral cortex is reduced by 13–17% following high Pb exposure.[107,108,125,155] Significant reductions following developmental Pb exposure have also been reported in maximal width of the hippocampus and its subregions, such as the CA1 region, dentate gyrus, and dentate molecular layer.[5,31,125] The reductions in hippocampal size are found in the absence of any brain weight differences,[5] suggesting that these effects at least cannot be explained by nonspecific or nutritional effects of Pb on general brain development.

6.2. Vascular Changes

Pentschew and Garro[150] described a disruption in the vascular development, evidenced by increased vascular permeability and swollen endothelial cells in the brain, following exposure to high levels of Pb during development. In the human, also, a dilatation of capillaries is seen with swelling of the endothelial cells.[149] It is generally assumed that Pb acts on the endothelial cells, causing an increased capillary permeability and edema, i.e., breakdown of the BBB[41,150,159,206]; however, this is not found at lower doses of Pb.[117]

Examination of the developing capillary system in high Pb-exposed animals revealed aberrations in the developing capillary buds, which were either dilated or lacked apical processes.[161] Electron microscopic examination has revealed capillary damage, with vessel walls containing numerous varicosities, the fusion of membranes periodically interrupted, hypertrophied capillaries and endothelial cells, and a few necrotic endothelial cells.[161,206] Lead appears to be concentrated, at least initially, in brain capillaries in animals exposed to Pb developmentally[209] and is localized in the cytoplasm of the capillary endothelium cells after an intraperitoneal (ip) injection of radioactive Pb.[207,224]

6.3. Effects on Myelin

Research on the effects of Pb on myelin have examined either its effects on myelin formation in young animals or segmental demyelination in adults.

In developing animals, Pb exposure produces a retardation in myelination. Biochemical analyses have shown a reduction of myelin basic protein, phospholipids, cholesterol, galactolipids, and sulfatides.[106,204] With the exception of myelin basic protein, which is reduced by 60%, most of the other myelin-rich chemicals are reduced by about 30–40%. Detailed biochemical analysis of the various myelin subfractions indicated that, although reduced, the myelin formed in Pb-exposed animals is indistinguishable from that of the normal rat.[106] Those authors also found a recovery from these deficiencies by 60 d of age, suggesting a capacity for compensation of myelin over time.

Morphological studies have revealed that there is reduced myelination in Pb-exposed developing rats.[106,204] Lead-exposed animals have fewer myelin lamellae than control animals. However, the reduced myelin is paralleled by a reduction in axonal size, suggesting that hypomyelination is secondary to reduced axonal maturation. There was no observed delay in the onset of myelination,[204] and ultrastructural examination suggested that the thickness of individual myelin lamella was not altered by Pb.[106] Brashear et al.[24] found that developmental Pb exposure was associated with initially swollen Schwann cells and a delayed myelin compaction, but the segmental demyelination characteristic of adult Pb exposure was not seen.

In adult animals the most consistent effect of Pb is segmental demyelination.[66,111,175,225] This effect is found in rats[111,225] and guinea pigs,[66] but not in humans and several other animals.[28,116] It is characterized by progressive demyelination and remyelination, but precedes the development of endoneurial edema.[224]

Thus, Pb appears to disrupt the normal formation of myelin when exposure occurs during development, and causes segmental demyelination when exposure occurs in adulthood. It is not known whether different species show different demyelinating effects of Pb, or whether these differences are the result of differences in methodology, exposure levels, or sensitivity.

6.4. Neuronal Changes

6.4.1. CELL NUMBER

Electron microscopic examination of neurons has been carried out by several researchers. Although some workers[107] observe no discernable differences in electron micrographs between Pb and control animals, others have reported the formation of lead-containing dense bodies in neuronal cytoplasm, frequently in cell processes.[89] At extremely high levels of Pb exposure incompatable with life, the following neuronal changes occur: extensive vacuolation, disorganization of the endoplasmic reticulum, folding of the nuclear membrane, swelling and severe lesions of the mitochondria, hypertrophy of golgi complexes, and patchy chromatolysis of the ergastoplasm.[67,143]

Examination of cell number by DNA analysis shows no change for the whole brain after 30 d high Pb exposure.[106] At 60 d of age, however, there was an increase in whole brain DNA, which the authors interpreted as diffuse gliosis in Pb-intoxicated animals. Lead does produce a 10–20% reduction in the DNA of the cerebellum in 3-week-old rat pups exposed to encephalopathic levels of Pb. This suggests that postnatally dividing cells of the cerebellum are sensitive to the effects of Pb, and that there may be an ultimate reduction in the number of cells in this brain area. Such a reduction in cerebellar DNA is not found after low doses of Pb.[199]

Neither Pentschew and Garro[150] nor Krigman et al.[109] were able to find any morphological evidence of a significant amount of cortical necrosis after developmental Pb exposure in the encephalopathic range.

Histological methods of cell number analysis reveal that there are more cells per area of the cerebral cortex (greater packing density) after Pb-exposure of developing rats.[107] However, when the reduced overall thickness of the cerebral cortex in the Pb-exposed animals is taken into account, the authors deduce that the number of neurons remains the same in Pb-exposed pups, but that they undergo less axodendritic development. Thus, other than in the cerebellum where cells of the external granular layer are undergoing cell division during the period of high level Pb exposure, no brain area has been shown to suffer a reduction in neuronal cell number. The hippocampus accumulates large amounts of Pb and is also experiencing rapid postnatal cellular development of the dentate gyrus granule cells at this time, but it has not been examined for reduced neuronal number.

6.4.2. DENDRITIC CHANGES

When examined electron microscopically, neuronal processes in Pb-exposed animals have a larger diameter, which is interpreted as the result of a restricted subdivision of the axons and dendrites.[108,109] Petit and LeBoutillier[155] examined the dendritic development of neocortical layer V pyramidal cells. In high Pb-exposed animals they found significant reductions in dendritic branches at distances greater than 20 μ from the cell body, and a 5.6% reduction in the length of the primary apical dendrite. Similarly, deficient dendritic development has been found in the cerebellum.[148] Alfano and Petit[5] examined dendritic development in the hippocampal dentate gyrus. They found the length of the dendritic field to be reduced, and alterations in dendritic branching consistent with a retardation in dendritic development. These reductions were seen in high-Pb animals as well as in animals exposed to levels of Pb that produce no encephalopathic signs.

6.4.3. AXONAL CHANGES

Peripheral nervous system axonal degeneration has been a consistent finding following Pb exposure in humans,[26] rhesus monkeys,[87] and guinea pigs.[66] Lead exposure in developing animals results in a reduction in the diameter of rat pyramidal tract axons.[106] Similar findings of reduced axonal size were observed in the optic nerves of Pb-exposed mice.[204]

A morphometric examination of the hippocampal mossy fiber system (the axons of the hippocampal dentate granule cells) following Pb exposure during development was carried out by Alfano et al.[3] The development of this axonal system is of particular interest since this pathway sequesters the largest amount of Pb in the normal brain, and forms a major internal circuit in the hippocampus. The authors report that high Pb exposure causes reductions in the development of this pathway, even in the absence of any differences in brain weights.

6.4.4. SYNAPTIC CHANGES

Four studies have examined synaptic development in the cerebral cortex of Pb-exposed rats. All four found a decrease in either the absolute number of synapses, or the number of synapses per neuron.[10,107,129,155] Though McCauley et al.[129] observed a less mature synaptic profile in their sample, Averill and Needleman[10] did not observe any differences with regard to synaptic maturity, and Krigman et al.[107] observed no maturational lag in the synapses in Pb-exposed pups in terms of either ganglioside composition or ultrastructural features, such as type of vesicles, area density of vesicles, surface of presynaptic membrane thickening, asymmetry of dense projections, and axial ratios of terminal boutons. Similarly, Petit and LeBoutillier[155] found no differences among developmentally Pb-exposed rats in any of their measurements of presynaptic length and thickness of the dense projection, postsynaptic length and thickness, or cleft width. Campbell et al.[31] have recently reported a decrease in complex synaptic terminals in the deep (close to the stratum pyramidale) part of the proximal (close to the dentate gyrus) region of the mossy fiber pathway in early Pb-exposed animals, which correlates with previously discussed alterations in the pathway. Further, those authors point out that those fibers originate from dentate granule cells located in the infrapyramidal limb, the same cells in which Alfano and Petit[5] found reduced dendritic development.

6.5. Anatomical Effects: Integration

Most of the evidence suggests that Pb retards dendritic and axonal development as well as the number synapses, but does not retard the maturation of the structural components of the synapse itself. The reduction in axonal development then induces a reduction in the formation of myelin. In the adult one sees a different picture, where there is segmental demyelination and axonal degeneration, neither of which occurs in the infant. This may be caused by the higher circulating levels of Pb tolerated by the adult; i.e., the infant might not tolerate or survive circulating Pb levels high enough to induce segmental demyelination.

7. Therapies in the Treatment of Lead-Induced Disorders

7.1. Chelation Studies

A number of natural substances, including sodium citrate and ascorbic acid, chelate Pb.[72] Commonly used chelating agents include calcium disodium edetate (CaEDTA), dimercaprol (BAL), 2,2-dimethylcysteine (D-penicillamine) and 2,3-dimercaptosuccinic acid (DMS).[65,74] Chisholm[38-40] and Coffin et al.[42] have outlined the appropriate protocols for chelation therapy for children with Pb poisoning, whereas Friedheim et al.[65] and the Center for Disease Control,[34,35] outline the treatment of adult Pb intoxication.

Pb appears to be bound in two different compartments, a relatively mobile and a relatively nonmobile, or stable, compartment. With the administration of chelators there is an initial rapid phase of Pb removal that may represent removal of extracellular Pb. This appears to shift the equilibrium concentration between cellular and extracellular compartments. Continued chelation then produces a slower phase of Pb elimination.[160] The Center for Disease Control[35] warns that "children who require chelation therapy will also require long-term medical surveillance and care. 'Rebound' of blood lead levels resulting from release of lead from tissue pools after an apparently successful course of chelation therapy should be anticipated." It is apparent, therefore, that although some tissues may have a very stable component of Pb (particularly bone), many tissues (including bone) also can have an active turnover of Pb. Graziano et al.[80] reported that DMS chelation causes 80% of the Pb administered to be excreted, predominantly in the urine.

Prior to the introduction of BAL in the late 1940s[203] and CaEDTA in the early 1950s,[96] the treatment of Pb poisoning had relied heavily on dietary measures aimed primarily at enhancing the deposition of Pb in bone.[124] The advent of chelation therapy dramatically reduced mortality from Pb encephalopathy from 65 to 25%.[120] In 1966, Coffin et al.[42] reported only one death among 22 children suffering from Pb encephalopathy using a combination of BAL and CaEDTA along with measures to control cerebral edema.

Several studies have attempted to assess whether chelation also results in reduced neurobehavioral consequences in children with increased Pb burdens. Sachs et al.[173] compared 47 children treated for Pb poisoning to their next in age siblings on a battery of psychological tests. Blood Pb levels in the treated group ranged from 50 to 365 μg/dL, with symptoms of Pb intoxication present in 18 and none demonstrating frank encephalopathy. In this investigation, intelligence tests indicated that there were no differences in IQ between children treated with chelators and their sibling controls, suggesting that chelation prevented intellectual decline. Pueschel et al.[157] reported an 8 point increase in the IQ of children treated with chelators. Alterations produced by chelators in children with very low levels of Pb are difficult to demonstrate.[50]

Chelation therapy, therefore, causes a dramatic reduction in the body burden of Pb and causes an equally dramatic reduction in the probability of mortality. Although some work has indicated the psychological benefits of chelation, insufficient research has been conducted on humans or experimental animals to allow a definitive statement about the efficacy of chelation therapy in improving neurobehavioral development following Pb exposure.

7.2. Therapeutic Animal Models

7.2.1. ENVIRONMENTAL ENRICHMENT

Environmental enrichment paradigms have frequently been used to demonstrate the plasticity of both the developing brain and some aspects of behavior. Major changes in brain development include increases in neocortical weight and thickness, increased dendritic branching and connectivity, and a greater number of neocortical synapses.[82,88,135,215,219] Behaviorally, superior problem-solving abilities have been demonstrated in enriched animals.[52,197]

Environmental enrichment has also been recently used as a therapy in the treatment of animals suffering from developmentally produced brain lesions[220,221] and hypothyroidism.[47] These studies reported improvement in the behavioral capacities of the enriched, brain-damaged animals.

Interestingly, many of the changes seen after environmental enrichment, e.g., increased neocortical development, are opposite to those observed in Pb-exposed animals. Petit and Alfano[153] examined the efficacy of environmental enrichment as a therapy for developmentally Pb-exposed animals. They observed that the enrichment paradigm was capable of reducing the passive avoidance deficit in low Pb exposed rats, but had no effect on the hyperactivity observed in high Pb exposed animals. Thus, the efficacy of environmental enrichment as a therapy for Pb-exposed animals may be both dose- and task-dependent.

7.2.2. PHARMACOLOGICAL MANIPULATIONS

Several investigators have evaluated the locomotor response of Pb-exposed animals to a number of psychoactive drugs, many of which are used in the diagnosis and treatment of hyperkinetic children.[218]

Silbergeld and Goldberg[190,191] initially reported that postnatally Pb-exposed mice responded paradoxically to the two stimulant drugs, amphetamine and methylphenidate. A subsequent investigation by the same research group[192] examined the locomotor response of Pb exposed mice to a variety of pharmacological agents including apomorphine, atropine, neostigmine, physostigmine, alpha-methylparatyrosine, benztropine, chlorpromazine, L-DOPA, phenfluramine, and 2-dimethyl-aminoethanol. The results of that study indicated that cholinergic antagonists and several aminergic agonist exacerbated Pb-induced hyperactivity, while cholinergic agonists and aminergic antagonists suppressed this hyperactivity. Rafales et al.[164] observed an altered response to d-amphetamine in Pb-exposed mice in the absence of any weight differences. Attenuated locomotor activity in developmentally Pb-exposed mice and rats following amphetamine administration has also been observed by several other authors.[79,101,200]

Taken together, these observations indicate that developmental exposure to Pb consistently alters the behavioral responsiveness of both rats and mice to amphetamine administration. The behavioral responsiveness to other psychoactive drugs had not yet been thoroughly and systematically evaluated.

The results to date, however, are consistent with neurochemical observations of increased aminergic and decreased cholinergic functioning in Pb-exposed animals. Those drugs that increase cholinergic functioning, counteracting Pb's cholinergic blocking effects, and those drugs that decrease aminergic functioning, counteracting Pb's aminergic enhancing effects, are effective as therapeutic agents. Although the implications of these findings for the potential psychopharmacological treatment of Pb exposure in humans are exciting, further clinical research is needed before definitive statements can be made.

8. Final Considerations

8.1. Methodological Considerations

8.1.1. METHODS AND LEVELS OF LEAD EXPOSURE

There have been several methods of Pb exposure in rodents developed since the classic Pentschew and Garro[150] model. The different methods of postnatal Pb exposure can be broken down into two major groups: (1) those in which Pb is administered indirectly through the mother (either in her food or water), which she then passes on to the infant through her milk, and (2) those in which a Pb-containing solution is administered directly to the infant, either through intubation[147] or injection.[25] There are advantages and disadvantages to each of the models, which probably accounts for the lack of generalized acceptance of any one model. The indirect method of Pb exposure generally involves placing lead carbonate or lead acetate in the food[156] or lead acetate in the drinking water.[191] Four percent lead carbonate or 5% lead acetate is generally the highest dose placed in the food, and at this dose most researchers have reported the development of an encephalopathy in the infants, usually between 18 and 25 d of age. Researchers thus generally employ these or lower doses with this treatment paradigm. Lead acetate in the drinking water is generally employed at doses of 1% or less and at the highest doses an encephalopathy also occurs. A concern with placing lead acetate in the drinking water is that it quickly precipitates out of the solution as lead carbonate. One method of counteracting this effect is to add acetic acid to the water, but this introduces an additional variable over Pb administration in the food. The problem can also be corrected by frequent water changes. The general problems with the indirect methods are (1) possible alterations in the mother's physiology (e.g., quality or quantity of milk), or behavior that may effect the infant (e.g., stress of eating Pb contaminated food, possible change in her behavior toward her infants), and (2) the infants are exposed to varying amounts of Pb, i.e., the amount consumed increases dramatically around postnatal day 16 when infants begin to consume the contaminated food or water themselves. These problems are generally overcome by limiting the size of the litter and/or using weight-matched litters to control for any potential malnutrition effects from changes in the mother's milk. The possibility of Pb altering maternal behavior is small, however, since ample research shows the adult animal (1) absorbs a minimal amount of Pb (see above) such that the mother's circulating Pb level is substantially lower than that of the infant, and (2) shows minimal behavioral changes following even high levels of Pb. Thus, it would appear that the only major drawback of the indirect exposure method is the inability to control accurately for exposure levels, and the most accepted way of countering this problem is by recording blood or brain Pb levels in the exposed animals.

The methods of direct exposure through intubation or injection attempt to rule out any effects exerted by maternal Pb consumption; however, this is not completely possible since the mother generally consumes the infant excretions, which would be laden with Pb, during its early development. Thus, the mothers are also being exposed to Pb since infant rats excrete up to 50% of their consumed Pb (see above). Another difficulty with the direct exposure procedures is the stress inflicted on the pups by daily intubation or injection. Stress itself is known to alter brain and behavioral development, thus this technique introduces another variable. One of the benefits of the intubation and injection techniques is that they can hold the dose of Pb constant for each animal, ensuring constant exposure levels. Therefore, the intubation/injection method controls for levels of Pb exposure over development, but introduces the variable of stress, which can make the results difficult to interpret.

Thus, neither the direct nor indirect methods of Pb exposure to developing animals is able to eliminate all potential confounding variables. In addition, the importance of some of the variables, such as constant exposure regimens, are difficult to assess since the rate of absorption of Pb from the gut and the ability of Pb to penetrate the brain vary greatly with the increasing age of the pup. Most researchers, however, are more concerned with the mechanisms of Pb action than the specific type of exposure regimen. Furthermore, most have come to rely on more direct methods of assessment of actual internal Pb through assessment of blood or brain Pb levels.

Bornschein et al.[18] presented data relating various methods of Pb exposure to actual Pb intake in mother and infant rats and mice. Eating a 4% lead carbonate diet, which contains 31,000 ppm Pb, rat dams are exposed to approximately 5 g Pb/kg body weight/d. Female rats transfer 10.1–14.2% of injected [203]Pb to suckling pups via lactation. Rat dams fed 4% lead carbonate produce 40 ppm Pb in their milk, and the neonate consumes about 1.0 mg/kg/d prior to weaning. When the young rats begin to eat the solid food around P16, their exposure goes up to 5 g/kg/d.

Mice dams exposed to a 0.5% lead acetate drinking solution consume approximately 1 g Pb/kg/d. They produce milk containing about 15 ppm Pb, which is less than half that seen in the above rat model. However, because of the high metabolic rate in the mouse, the higher intake of milk per body weight yields a daily Pb exposure of approximately 1 mg/kg/d, very similar to the rat model. When the young mice begin to drink the Pb-contaminated water, their exposure jumps to 0.6–1 g/kg/d.

8.1.2. COMPARATIVE ASPECTS OF LEAD NEUROTOXICITY

Much of the difficulty in conducting research on humans can be reduced by using animal models. However, for reasons listed above, a universally suitable animal model has not been chosen.

One of the misconceptions often found in animal research is reflected in attempts to produce blood Pb levels in rodents in the 40–80 μg/dL range in order to correspond to those seen in the human. Petit and Alfano[153] and Goldberg[69] have pointed out that the rodent has a much greater tolerance for Pb than the human, and consequently has a different dose–response curve. As Goldberg[69] states, "Rodents with blood levels exceeding 100 μg/dL do not present symptoms of overt intoxication, while man with the same blood level of Pb is clearly intoxicated." Encephalopathy and death, which occur in children at approximately 200–300 μg/dL[81,149] do not occur in the rat until blood Pb levels reach approximately 1000 μg/dL.[153] At 300 μg/dL, the rat shows no weight loss or other overt neuropathological signs, although behavioral alterations have been reported (see Petit and Alfano[153] and Goldberg[69] for further discussion). Therefore, research workers conducting studies in rodents cannot use blood Pb levels to compare their results directly to those in humans with the same blood

Pb levels. While further research is clearly needed to clarify this issue, it would appear at present that rodent blood Pb levels must be divided by at least three or four to approximate the comparable human condition.

Another interesting observation in research with Pb-treated animals is the difficulty in replicating results between laboratories. Much of this can be explained as caused by the methodological differences previously discussed, as well as by differences in levels of Pb exposure. However, an increased variability in Pb-exposed rats is usually apparent. That is, some animals, or even litters, appear to be more adversely affected by Pb than others. The result is a greater variability in the Pb groups than in the control groups, with some animals showing Pb effects and others being more resistant. This phenomenon has increased the difficulty in understanding Pb effects when routine statistical methods of analysis are used.

A final problem found of in vitro research is that physiological solutions of Pb are extremely difficult to prepare and maintain, since Pb causes a precipitate in most solutions. This presents a problem in general bathing solutions, and is a particular problem when attempting to eject Pb solutions from micropipets, where small Pb precipates quickly clog the tip.

8.2. Conclusions

A great deal of research has been generated in recent years in trying to understand the neurobiological and behavioral effects of Pb. Although much is yet unknown, particularly at the human level, several points concerning the mechanisms of Pb action are becoming clear. Lead appears to be a highly toxic substance that is more readily absorbed by, and more easily enters the brain in young animals where it causes immediate effects by altering chemistry and physiology. These functional alterations may result in morphological alterations in the brain that can remain after Pb levels have fallen, and induce long-term behavioral dysfunctions.

Research interest in Pb is generated at several levels. Many workers are interested in understanding and preventing adverse effects in humans. Perhaps the most hopeful area for applying the results of basic Pb research is the possible use of psychopharmacological therapies, developed in animals, to treat Pb-exposed humans, particularly children. However, beyond the human implications, many workers have become interested in the effect of Pb as a neurobiological tool for studies of CNS function (e.g., as a Ca blocker). Others have become interested in Pb because of the increasing incidence of Pb toxicity in waterfowl consuming Pb shot.

It has long been clear that Pb is a neurobehavioral toxin; what is now needed is more information about how it exerts its neurobehavioral effects and how to counteract them. Finally, society as a whole should continue taking steps to reduce the widespread contamination of our environment by lead.

Acknowledgment

Preparation of this manuscript, as well as all research cited from our laboratory, was supported by Grant No. A0292 and G0165 from the Natural Sciences and Engineering Research Council of Canada to T.L.P.

References

1. Albert, R. E., Shore R. E., Sayers, A. J. Strehlow, C., Kneip, T. J., Pasternack, B. S., Friedhoff, A. J., Covan, F. and Cimino, J. A. (1974). Follow-up of children overexposed to lead. *Environ. Hlth. Persp.* 7, 33–39.
2. Alexander, F. W., Delves, H. T. and Clayton, B. E. (1973). The uptake and excretion by children of Pb and other contaminants. In: *Proceedings of the International Symposium Environmental Health Perspectives on Lead*, Amsterdam, 2-6 October, 1972, Commission of European Communities, Luxembourg, pp. 319–331.
3. Alfano, D. P., LeBoutillier, J. C. and Petit, T. L. (1982). Hippocampal mossy fiber pathway development in normal and postnatally lead exposed rats. *Exp. Neurol.* 75, 308–319.
4. Alfano, D. P., and Petit, T. L. (1981). Behavioral effects of postnatal lead exposure: Possible relationship to hippocampal dysfunction. *Behav. Neur. Biol.* 32, 319–33.
5. Alfano, D. P. and Petit, T. L. (1982). Neonatal lead exposure alters the dendritic development of hippocampal dentate granule cells. *Exp. Neurol.* 75, 275–288.
6. Allen, J. R., McWey, P. J. and Suomi, S. J. (1974). Pathobiological and behavioral effects of lead intoxication in the infant rhesus monkey. *Environ. Hlth. Persp.* 7, 239–246.
7. American Academy of Pediatrics (1969). Prevention, diagnosis, and treatment of lead poisoning in children. *Pediat.* 44, 291–298.
8. Araki, S. and Honma, T. (1976) Relationships between Pb absorption and peripheral nerve conduction velocities in lead workers. *Scand. J. Work Environ. Hlth.* 4, 225–231.

9. Arnvig, E., Grandjean, P., and Beckmann, J. (1980). Neurotoxic effects of heavy lead exposure determined with psychological tests. *Toxicol. Lett.* **5**, 399–404.

10. Averill, D. R. and Needleman, H. L. (1980). Neonatal lead exposure retards cortical synapatogenesis in the rat. In: H. L. Needleman (Ed.), *Low Level Lead Exposure: The Clinical Implications of Current Research*, Raven, New York, pp. 201–210.

11. Baloh, R., Sturm, R., Green, B. and Gleser, G. (1975). Neuropsychological effects of chronic asymptomatic increased lead absorption. *Arch. Neurol.* **32**, 326–330.

12. Barbeau, A. (1973). The biochemistry of Huntingtons Chorea: Recent developments. *Psychiat. Forum* **3**, 8–12.

13. Barltrop, D. and Khoo, H. E. (1975) Nutritional determinants of lead absorption. In: D. D. Hemphill (Ed.), *Trace Substances in Environmental Health—IX*, University of Missouri, Columbia, pp. 369–376.

14. Barry, P. S. I. and Mossman, D. B. (1970). Lead concentrations in human tissues. *Brit. J. Ind. Med.* **27**, 339–351.

15. Barton, J. C., Conrad, M. E., Harrison, L. and Nuby, S. (1978). Effects of calcium on the absorption and retention of lead. *J. Lab. Clin. Med.* **91**, 366–376.

16. Bloomfield, J. J., Trasko, V. M., Sayers, R. R., Page, R. T. and Peyton, M. F. (1940). A Preliminary Survey of the Industrial Hygiene Problem in the United States. DHEW Public Health Service Bulletin No. 259. Washington.

17. Bondy, S. C., Anderson, C. L., Harrington, M. E. and Prasad, K. N. (1979). The effects of lead and mercury on neurotransmitter high-affinity transport and release mechanisms. *Environ. Res.* **18**, 102–111.

18. Bornschein, R. L., Michaelson, I. A., Fox, D. A. and Loch, R. (1977). Evaluation of animal models used to study effects of Pb on neurochemistry and behavior. In: S. D. Lee (Ed.), *Biochemical Effects of Environmental Pollutants*, Ann Arbor Science, Ann Arbor pp. 441–460.

19. Bornschein, R., Pearson, D. and Reiter, L. (1980a). Behavioral effects of moderate lead exposure in children and animal models: Part 1, Clinical studies. *CRC Crit. Rev. Toxicol.* **8**, 43–99.

20. Bornschein, R., Pearson, D., and Reiter, L. (1980b). Behavioral effects of moderate lead exposure in children and animal models: Part 2, Animal studies. *CRC Crit. Rev. Toxicol.* **8**, 101–152.

21. Bornstein, J. C. and Pickett, J. B. (1977) Some effects of lead ions on transmitter release at rat neuromuscular junctions. *Soc. Neurosci. Abst.* **3**, 370.

22. Bradley, J. E. and Powell, A. M. (1954) Oral calcium EDTA in lead intoxication of children. *J. Pediat.* **45**, 297–301.

23. Brady, K., Herrera, Y. and Zenick, H. (1975). Influence of parental lead exposure on subsequent learning ability of offspring. *Pharmol. Biochem. Behav.* **3**, 561–565.

24. Brashear, C. W., Kopp, V. J. and Krigman, M. R. (1978). Effect of lead on the developing peripheral nervous system. *J. Neuropath. Exp. Neurol.* **38**, 414–425.

25. Brown, D. R. (1975). Neonatal lead exposure in the rat: Decreased learning as a function of age and blood lead concentrations. *Toxicol. Appl. Pharmacol.* **32**, 628–637.

26. Buchthal, F. and Behse, F. (1979). Electrophysiology and nerve biopsy in men exposed to lead. *Br. J. Ind. Med.* **36**, 135–147.

27. Burrows, N. F. E., Rendle-Short, J. and Hanna, D. (1951). Lead poisoning in children. *Brit. Med. J.* **1**, 329–334.

28. Bushnell, P. J. and Bowman, R. E. (1979). Reversal learning deficits in young monkeys exposed to lead. *Pharmacol. Biochem. Behav.* **10**, 733–742.

29. Byers, R. K., and Lord, E. E. (1943). Late effects of lead poisoning on mental development. *Amer. J. Dis. Child.* **66**, 471–494.

30. Byers, R. K. and Maloof, C. (1954). Edathamil calcium disodium (versenate) in treatment of lead poisoning in children. *Amer. J. Dis. Child.* **87**, 559–569.

31. Cambell, J. B., Wooley, D. E., Vijayan, V. K. and Overmann, R. (1982). Morphometric effects of postnatal lead exposure on hippocampal development of the 15 day old rat. *Dev. Brain Res.* **3**, 595–612.

32. Carroll, P. T. Silbergeld, E. K. and Goldberg, A. M. (1977). Alterations in central cholinergic function in lead induced hyperactivity. *Biochem. Pharmacol.* **26**, 397–402.

33. Catton, M. J., Harrison, M. J. G., Fullerton, P. M. and Kazantzis, G. (1970). Subclinical neuropathy in lead workers. *Br. Med. J.* **2**, 80–82.

34. Center for Disease Control (1975). Increased lead absorption and lead poisoning in young children. *J. Pediat.* **87**, 824–830.

35. Center for Disease Control (1978). Preventing lead poisoning in young children. DHEW Public Health Service Publication No. 2629, Washington.

36. Cerklewski, F. L. and Forbes, R. M. (1976). Influences of dietary zinc on lead toxicity in the rat. *J. Nutr.* **106**, 689–696.

37. Chang, L. W., Wade, P. R., Pounds, J. G. and Reuhl, K. R. (1980). Prenatal and neonatal toxicology and pathology of heavy metals. *Adv. Pharmacol. Chemother.* **17**, 195–231.

38. Chisolm, J. J. (1968). The use of chelating agents in the treatment of acute and chronic lead intoxication in childhood. *J. Pediat.* **73**, 1–38.

39. Chisolm, J. J. (1971). Treatment of lead poisoning. *Modern Treatment* **8**, 593–611.

40. Chisolm, J. J., and Barltrop, D. (1979). Recognition and management of children with increased lead absorption. *Arch. Dis. Child.* **54**, 249–262.

41. Clasen, R. A., Hartman, J. F., Coogan, P. S., Pandolfi, S., Laing, I. and Becker, R. A. (1974). Experimental acute lead encephalopathy in the juvenile rhesus monkey. *Expt. Hlth. Perspect.* **7**, 175–185.

42. Coffin, R., Phillips, J. L., Staples, W. I. and Spector, S. (1966). Treatment of lead encephalopathy in children. *J. Pediat.* **69**, 198–206.

43. Cooper, G. P. and Steinberg, D. (1977). Effects of cadmium and lead on adrenergic neuromuscular transmission in the rabbit. *Amer. J. Physiol.* **232**, 128–131.

44. Culver, B. W. (1976). Epidemiological considerations of occupational lead exposure. In: B. W. Carnow (Ed.), Health Effects of Occupational Lead and Arsenic Exposure. DHEW Public Health Service Publication No. 76-134. Washington.

45. Dagg, J. H., Goldberg, A., Lochhead, A. and Smith, J. H. (1965). The relationship of lead poisoning to acute intermittent porphyria. *Quart. J. Med.* **34**, 163–175.

46. Danscher, G., Fjerdingstad, E. J., Fjerdingstad, E. and Fredens, K. (1976). Heavy metal content in subdivisions of the rat hippocampus (zinc, lead and copper). *Brain Res.* **112**, 442–446.

47. Davenport, J. W. (1976). Environmental Therapy in Hypothyroid and Other Disadvantaged Animal Populations. *Adv. Behav. Biol.* **17**, Plenum, New York.

48. David, O. J. (1974). Association between lower level lead concentrations and hyperactivity in children. *Environ. Hlth. Persp.* **7**, 17–25.

49. David, O. J., Clark, J. and Voeller, K. (1972). Lead and hyperactivity. *Lancet* **2**, 900–903.

50. David, O. J., McCann, B., Sverd, J. and Clark, J. (1976). Low lead levels and mental retardation. *Lancet* **2**, 1376–1379.

51. De la Burde, B. and Choate, M. S. (1972). Does asymptomatic lead exposure in children have latent sequalae? *J. Pediat.* **81**, 1088–1091.

52. Dennenberg, V. H., Woodcock, J. M. and Rosenberg, K. M. (1968). Long term effects of preweanling and postweanling free environment experience on rat's problem solving behavior. *J. Comp. Physiol. Psychol.* **66**, 533–535.

53. Driscoll, J. W. and Stegner, S. E. (1976a). Behavioral effects of chronic lead ingestion on laboratory rats. *Pharmacol. Biochem. Behav.* **4**, 411–417.

54. Driscoll, J. W. and Stegner, S. E. (1976b). Lead produced changes in the relative rate of open field activity of laboratory rats. *Pharmacol Biochem. Behav.* **8**, 743–747.

55. Dubas, T. C. and Hrdina, P. D. (1978). Behavioral and neurochemical consequences of neonatal exposure to lead in rats. *J. Enrir. Pathol. Toxicol.* **2**, 473–484.

56. Environmental protection Agency (1977). Air quality criteria for lead. EPA Publication No. 600/8-77-017. Washington.

57. Fejerman, N., Gimenez, E., Vallejo, N. and Medina, C. (1973). Lennox's syndrome and lead intoxication. *Pediat.* **52**, 227–234.

58. Feldman, R. G., Haddow, J. and Chisolm, J. J. (1972). Chronic lead intoxication in urban children. In: J. Desmedt and S. Karger (Eds.), *New Developments in Electromyography and Clinical Neurophysiology* **2**, 313–317.

59. Feldman, R. G., Hayes, M. K., Younes, R. and Aldrich, F. D. (1977). Lead neuropathy in adults and children. *Arch. Neurol.* **34**, 481–488.
60. Fjerdingstad, E. J., Danscher, G. and Fjerdingstad, E. (1974). Hippocampus: selective concentration of lead in the normal rat brain. *Brain Res.* **80**, 350–354.
61. Forbes, G. B. and Reina, J. C. (1972). Effect of age on gastrointestinal absorption (Fe, Sr, Pb) in the rat. *J. Nutr.* **102**, 647–652.
62. Fox, D. A., Lewkowski, J. P. and Cooper, G. P. (1977). Acute and chronic effects of neonatal lead exposure on development of the visual evoked response in rats. *Toxicol. App. Pharmacol.* **40**, 449–461.
63. Fox, D. A., Lewkowski, J. P. and Cooper, G. P. (1979) Persistent visual cortex excitability alterations produced by neonatal lead exposure. *Neurobehav. Toxicol.* **1**, 101–106.
64. Fox, D. A., Overmann, S. R. and Wooley, D. E. (1979). Neurobehavioral ontogeny of neonatal lead-exposed rats. II. Maximal electroshock seizures in developing and adult rats. *Neurotoxicol.* **1**, 149–170.
65. Friedheim, E., Graziano, J. H., Popovac, D., Dragovic, D. and Kaul, B. (1978). Treatment of lead poisoning by 2,3-dimercaptosuccinic acid. *Lancet* **2**, 1234–1236.
66. Fullerton, P.M. (1966). Chronic peripheral neuropathy produced by lead poisoning in guinea pigs. *J. Neuropath. Exptl. Neurol.* **25**, 214–236.
67. Gennaro, L. D. (1978). The effects of lead nitrate on the central nervous system of the chick embryo. I. Observations of light and electron microscopy. *Growth*, **42**, 141–155.
68. Goldberg, A. M. (1977). Neurotransmitter mechanisms in inorganic lead poisoning. In: S. D. Lee (Ed.), *Biochemical Effects of Environmental Pollutants*, Ann Arbor Science, Ann Arbor, pp. 413–423.
69. Goldberg, A. M. (1979). The effects of inorganic lead on cholinergic transmission. *Prog. Brain Res.* **49**, 465–470.
70. Goldstein, G. W. and Diamond, T. (1973). Metabolic basis of lead encephalopathy. In: F. Plum (Ed.), *Brain Dysfunction and Metabolic Disorders*, Raven, New York, pp. 293–304.
71. Golter, M. and Michaelson, I. A. (1975). Growth, behavior and brain catecholamines in lead exposed neonatal rats: A reappraisal. *Science* **187**, 359–361.
72. Goyer, R. A. and Cherian, M. G. (1979). Ascorbic acid and EDTA treatment of lead toxicity in rats. *Life Sci.* **24**, 433–438.
73. Goyer, R. A. and Rhyne, B. C. (1973). Pathological effects of lead. *Int. Rev. Exp. Pathol.* **12**, 1–77.
74. Graef, J. W. (1980). Management of low level lead exposure. In: H. L. Needleman (Ed.), *Low Level Lead Exposure: The Clinical Implications of Current Research*. Raven, New York, pp. 121–126.
75. Grandjean, P. (1978a). Occupational lead exposure in Denmark: Screening with the hematofluorometer. *Brit. J. Ind. Med.* **36**, 52–58.
76. Grandjean P. (1978b). Widening perspectives of lead toxicity: A review of health effects of lead exposure in adults. *Env. Res.* **17**, 303–321.

77. Grandjean, P. (1978c). Regional distribution of lead in human brains. *Toxicol. Lett.* **2**, 65–69.

78. Grandjean, P., Arnvig, E. and Beckmann, J. (1978). Psychological dysfunctions in lead exposed workers. Relation to biological parameters of exposure. *Scand. J. Work Env. Hlth.* **4**, 295–303.

79. Gray, L. E. and Reiter, L. W. (1977). Lead-induced developmental and behavioral changes in the mouse. *Toxicol. Appl. Pharmacol.* **41**, 140.

80. Graziano, J. H., Leong, J. K., and Friedheim, E. (1978). 2,3-Dimercaptosuccinic acid: A new agent for the treatment of lead poisoning. *J. Pharmacol. Exp. Therapeut.* **206**, 696–700.

81. Greengard, J., Adams, B., and Berman, E. (1965). Acute lead encephalopathy in young children. *J. Pediat.* **66**, 707–711.

82. Greenough, W. T. and Volkmar, F. R. (1973). Pattern of dendritic branching in occipital cortex of rats reared in complex environments. *Exp. Neurol.* **40**, 491–504.

83. Hanninen, H., Mantere, P., Hernberg, S., Seppalainen, A. M. and Kock, B. (1979). Subjective symptoms in low-level exposure to lead. *Neurotoxicol.* **1**, 333–347.

84. Hansen, J. C., Christensen, L. B. and Tarp, U. (1980). Hair lead concentration in children with minimal cerebral dysfunction. *Dan. Med. Bull.* **27**, 259–262.

85. Hastings, L., Cooper, G. P., Bornschein, R. L. and Michaelson, I. A. (1979). Behavioral deficits in adult rats following neonatal lead exposure. *Neurobehav. Toxicol.* **1**. 227–231.

86. Hertz, M. M., Bolwig, T. G., Grandjean, P. and Westergaard, E. (1981). Lead poisoning and the blood-brain barrier. *Acta Neurol. Scand.* **63**, 286–296.

87. Heywood, R., James, R. W., Sortwell, R. J., Prentice, D. E. and Barry, P. S. I. (1978). The intravenous toxicity of tetra-alkyl lead compounds in rhesus monkeys. *Toxicol. Lett.* **2**, 187–197.

88. Holloway, R. L. (1966). Dendritic branching: Some preliminary results of training and complexity in rat visual cortex. *Brain Res.* **2**, 393–396.

89. Holtzman, D., Herman, M. M., Shen-Hsu, J. and Mortell, P. (1980). The pathogenesis of lead encephalopathy. *Virchows Arch. A. Path. Anat. Histol.* **387**, 147–164.

90. Hrdina, P. D., Hanin, I. and Dubas, T. C. (1979). Neurochemical correlates of lead toxicity. In: R. L. Singhal and J. A. Thomas (Eds.), *Lead Toxicity*. Urban and Schwarzenberg, Baltimore, pp. 273–300.

91. Hrdina, P. D., Peters, D. A. V. and Singhal, R. L. (1976). Effects of chronic exposure to cadmium, lead, and mercury on brain biogenic amines in the rat. *Res. Comm. Chem. Pathol. Pharmacol.* **15**, 483–493.

92. Hursh, J. B. and Suomela, J. (1968). Absorption of Pb from the gastrointestinal tract of man. *Acta Radiol.* **7**, 108–120.

93. Irwig, L. M., Harrison, W. O., Rocks, P., Webster, I. and Andrew M. (1978). Lead and morbidity: A dose response relationship. *Lancet* **2**, 4–7.

94. Jason, K. M. and Kellogg, C. K. (1980). Behavioral neurotoxicity of lead. In: R. L. Singhal and J. A. Thomas (Eds.), *Lead Toxicity*, Urban and Schwarzenberg, Baltimore, pp. 241–271.

95. Jones, D. L. (1979). Effects of neonatal lead exposure on spontaneous alternation performance in rats. *Soc. Neurosci. Abst.* No. 2214.

96. Karpinski, F. E., Raiders, F. and Girsh, L. S. (1953). Calcium disodium versenate in the therapy of lead encephalopathy. *J. Pediat.* **42**, 687–699.

97. Kehoe, R. A. (1961). The metabolism of lead in health and disease. *Arch. Environ. Hlth.* **2**, 418–422.

98. Klein, A. W. and Koch, T. R. (1981). Lead accumulations in brain, blood and liver after low dosing in neonatal rats. *Arch. Toxicol.* **47**, 257–262.

99. Kober, T. E. and Cooper, G. P. (1975). Competitive action of lead and calcium on transmitter release in bullfrog sympathetic ganglia. *Fed. Prod.* **34**, 404.

100. Kober, T. E. and Cooper, G. P. (1976). Lead competitively inhibits calcium dependent synaptic transmission in the bullfrog sympathetic ganglion. *Nature* **262**, 704–705.

101. Kostas, J., McFarland, D. J. and Drew, W. G. (1978). Lead induced behavioral disorders in the rat: Effects of amphetamine. *Pharmacol.* **16**, 226–236.

102. Kostial, K., Simonovic, I. and Pisonic, M. (1971). Lead absorption from the intestine of newborn rats. Nature, **233**, 564.

103. Kostial, K. and Vouk, V. B. (1957). Lead ions and synaptic transmission in the superior cervical ganglion of the cat. *Brit. J. Pharmacol.* **12**, 219–222.

104. Kotok. D., Kotok, R. and Heriot, J. T. (1977). Cognitive evaluation of children with elevated blood lead levels. *Amer. J. Dis. Child.* **131**, 791–793.

105. Krehbiel, D., Davis, G. A., LeRoy, L. M. and Bowman, R. E. (1976). Absence of hyperactivity in lead exposed developing rats. *Environ. Hlth. Perspec.* **18**, 147–157.

106. Krigman, M. R., Druse, M. J., Traylor, T. D., Wilson, M. H., Newell, L. R., and Hogan, E. L. (1974a). Lead encephalopathy in the developing rat: effect on myelination. *J. Neuropath. Exptl. Neurol.* **33**, 58–73.

107. Krigman, M. R., Druse, M. J., Traylor, T. D., Wilson, M. H., Newell, L. R., and Hogan, E. L. (1974b). Lead encephalopathy in the developing rat: effect on cortical ontogenesis. *J. Neuropath. Exptl. Neurol.* **33**, 671–686.

108. Krigman, M. R. and Hogan, E. L. (1974). Effect of lead intoxication on the postnatal growth of the rat nervous system. *Environ. Hlth. Perspect.* **7**, 187–199.

109. Krigman, M. R., Traylor, D. T., Hogan, E. L. and Mushak, P. (1974). Ninety-nanometer dense core vesicles: A normal organelle in certain central nervous system myelinated axons of the guinea pig. *J. Neuropath. Exp. Neurol.* **33**, 561–562.

110. Lampert, P., Garro, F. and Pentschew, A. (1967). Lead encephalopathy in suckling rats: An electron microscopic study. In: I. Klatzo and F. Seitelberger (Eds.), *Symposium on Brian Edema* Springer, New York, pp. 207–222.

111. Lampert, P. W. and Schochet, S. S. (1968). Demyelination and remyelination in lead neuropathy. *J. Neuropath. Exp. Neurol.* **27**, 527–544.

112. Landrigan, P. J., Baker, E. L., Whitworth, R. H. and Feldman, R. G. (1980). Neuroepidemiologic evaluations of children with chronic increased lead absorption. In H. L. Needleman (Ed.), *Low Level Lead Exposure: The Clinical Implications of Current Research*, Raven, New York, pp. 17–33.

113. Landrigan, P. J., Baker, E. L., Jr., Feldman, R. G., Cox, D. H., Eden, K. V., Orenstein, W. A., Nather, J. A., Yankel, A. J. and Von Lendern, I. H. (1976). Increased lead absorption with anemia and slowed nerve conduction in children near a lead smelter. *J. Pediat.* **89**, 904–910.

114. Landrigan, P. J., Whitworth, R. H., Boloh, R. W., Staehling, N. W., Barthel, W. F. and Rosenblum, B. F. (1975). Neuropsychological dysfunction in children with chronic low-level lead absorption. *Lancet* **2**, 708–712.

115. Lane, R. E., Hunter, D., Malcolm, D., Williams, M. K., Hudson, T. G. F., Browne, R. C., McCallum, R. I., Thompson, A. R., deKretser, A. J., Zielhuis, R. L., Cramer, K., Barry, P. S. I., Goldberg, A., Beritic, T., Vigliani, E. C., Truhaut, R., Kehoe, R. A. and King, E. (1968). Diagnosis of inorganic lead poisoning: A statement. *Brit. Med. J.* **4**, 501.

116. Lee, W. R. (1981). What happens in lead poisoning. *J. Royal Coll. Physic. London* **15**, 48–54.

117. Lefauconnier, J. M., Lavielle, E., Terrien, N., Bernard, G. and Fournier, E. (1980). Effect of various lead doses on some cerebral capillary functions in the suckling rat. *Toxicol. Appl. Pharmacol.* **55**, 467–476.

118. Lilis, R., Fischbein, A., Diamond, S., Anderson, H. A., Selikoff, I. J., Blumberg, W. E. and Eisinger, J. (1977). Lead effects among secondary lead smelter workers with blood lead levels below 80 µg/10 mL. *Arch. Environ. Hlth.* **38**, 256–266.

119. Lilis, K., Fischbein, A., Eisinger, J., Blumberg, W. E., Diamond, S., Anderson, H. A., Rom, W., Rice, C., Sarkozi, L., Kon, S. and Selikoff, I. J. (1977). Prevalence of lead disease among secondary lead smelter workers and biological indicators of lead exposure. *Environ. Res.* **14**, 255–285.

120. Lin-Fu, J. S. (1967). Lead poisoning in children. DHEW. Public Health Service Publication No. 78-5142. Washington.

121. Lin-Fu, J. S. (1970). Lead poisoning in children. DHEW Public Health Service Publication No. 2108. Washington.

122. Lin-Fu, J. S. (1975). Undue lead absorption and lead poisoning in children—An overview. In: *Proceedings of the International Confer-*

ence on Heavy Metals in the Environment. Vol. 3., U. Toronto Press, Toronto, pp. 29–52.

123. Lin-Fu, J. S. (1979). Lead exposure among children—A reassessment. *N. Engl. J. Med.* **300**, 731–732.

124. Lin-Fu, J. S. (1980). Lead poisoning and undue lead exposure in children: History and current status. In H. L. Needleman (Ed.), *Low Level Lead Exposure: The Clinical Implications of Current Research,* Raven, New York, pp. 5–16.

125. Louis-Ferdinand, R. T., Brown, D. R., Fiddler, S. F., Daughtrey, W. C. and Klein, A. W. (1978). Morphometric and enzymatic effects of neonatal lead exposure in the rat brain. *Toxicol. Appl. Pharmacol.* **43**, 531–560.

126. Mahaffey, K. R. and Michaelson, I. A. (1980). The interaction between lead and nutrition. In: H. L. Needleman (Ed.), *Low Level Lead Exposure: The Clinical Implications of Current Research,* Raven, New York, pp. 159—199.

127. Mahaffey-Six, K. and Goyer, R. A. (1970). Experimental enhancement of lead toxicity by low dietary calcium. *J. Lab. Clin. Med.* **76**, 933–942.

128. Manalis, R. S. and Cooper, G. P. (1973). Presynaptic and postsynaptic effects of lead at the frog neuromuscular junction. *Nature* (London) **243**, 354–355.

129. McCauley, P. T., Bull, R. J. and Lutkenhoff, S. D. (1979). Association of alterations in energy metabolism with lead induced delays in rat cerebral cortical development. *Neuropharm.* **18**, 93–101.

130. Meredith, P. A., Moore, M. R., and Goldberg, A. (1977). The effect of calcium on lead absorption in rats. *Biochem. J.* **166**, 531–537.

131. Michaelson, I. A., Greenland, R. D. and Roth, W. (1974). Increased brain norepinephrine turnover in lead exposed hyperactive rats. *Pharmacologist* **16**, 250.

132. Michaelson, I. A., and Sauerhoff, M. W. (1974). Animal models of human disease: Severe and mild encephalopathy in the neonatal rat. *Environ. Hlth. Persp.* **7**, 201–225.

133. Modak, A. T., Weintraub, S. T. and Stavinoha, W. B. (1975a). Effects of chronic ingestion of lead on the central cholinergic system in rat brain regions. *Toxicol. Appl. Pharmacol.* **34**, 340–347.

134. Modak, A. T., Weintraub, S. T. and Stavinoha, W. B. (1975b). Chronic lead ingestion and the cholinergic system. *Pharmacol.* **17**, 213.

135. Mollgaard, K., Diamond, M. C., Bennet, E. L., Rosenzweig, M. R. and Lindner, B. (1971). Quantitative synaptic changes with differential experience in rat brain. *Int. J. Neurosci.* **2**, 113–128.

136. Morris, C. E., Heyman, A. and Pozefsky, T. (1964). Lead encephalopathy caused by ingestion of illicitly distilled whiskey. *Neurol.* **14**, 493–499.

137. National Academy of Sciences (1972). Lead: Airborne lead in perspective. NAS Publication, Washington, pp. 330–381.

138. National Bureau of Standards (1976). Survey manual for estimating the incidence of lead paint in housing. NBS Technical Note No. 921. Washington.

139. Needleman, H. L. (1973). Lead poisoning in children: Neurologic implications of widespread subclinical intoxication. *Semin. Psychiat.* 5, 47–54.

140. Needleman, H. L. (1977). Studies in subclinical lead exposure. *Envir. Hlth. Res. Ser.* June 1977, PB-271649. Health Effects Research Laboratories, USEPA, Research Triangle Park, North Carolina.

141. Needleman, H. L. (1980). Lead and neuropsychological deficit: Finding a threshold. In H. L. Needleman (Ed.), *Low Level Lead Exposure: The Clinical Implications of Current Research*, Raven, New York, pp. 43–51.

142. Needleman, H. L., Gunnoe, C., Leviton, A., Reed, R., Peresie, H., Maher, C. and Barrett, P. (1979). Deficits in psychologic and classroom performance of children with elevated dentine lead levels. *N. Engl. J. Med.* 300, 689–695.

143. Niklowitz, W. J. (1974). Ultrastructural effects of acute tetraethyllead poisoning on nerve cells of the rabbit brain. *Environ. Res.* 8, 17–36.

144. Niklowitz, W. J. and Yeager, D. W. (1973). The interference of Pb with essential brain tissue Cu, Fe, and Zn as a main determinant in experimental tetraethyllead encephalopathy. *Life Sci.* 13, 897–905.

145. Otto, D. A., Benignus, V. A., Muller, K. E. and Barton, C. N. (1981). Effects of age and body lead burdens on CNS function in young children. I. Slow cortical potentials. *EEG Clin. Neurophysiol.* 52, 229–239.

146. Overmann, S. R. (1977). Behavioral effects of asymptomatic lead exposure during neonatal development in rats. *Toxicol. Appl. Pharmacol.* 41, 459–471.

147. Overmann, S. R., Fox, D. A. and Wooley, D. E. (1979). Neurobehavioral ontogeny of neonatally lead exposed rats. I. Reflex development and somatic indices. *Neurotoxicol.* 1, 125–147.

148. Patrick, G. W., Anderson, W. J., and Brophy, P. D. (1979). Purkinje cell dendritic development in lead treated cats and rats. *Soc. Neurosci. Abst.* No. 562.

149. Pentschew, A. (1965). Morphology and morphogenesis of lead encephalopathy. *Acta Neuropathol.* 5, 133–160.

150. Pentschew, A. and Garro, F. (1966). Lead encephalo-myelopathy of the suckling rat and its implications on the porphyrinopathic nervous diseases, with special reference to the permeability disorders of the nervous system capillaries. *Acta Neuropathol.* 6, 266–278.

151. Perino, J. and Ernhart, C. B. (1974). The relation of subclinical lead level to cognitive and sensorimotor impairment in black preschoolers. *J. Learning Disabil.* 7, 26–30.

152. Perlstein, M. A. and Attala, R. (1966). Neurologic sequelae of plumbism in children. *Clin. Pediat.* 5, 292–297.

153. Petit, T. L. and Alfano, D. P. (1979). Differential experience following developmental lead exposure: Effects on brain and behavior. *Pharmacol. Biochem. Behav.* 11, 156–171.

154. Petit, T. L., Alfano, D. P. and LeBoutillier, J. C. (1983). Early lead exposure and the hippocampus. *Neurotoxicol.*, in press.
155. Petit, T. L. and LeBoutillier, J. C. (1979). The effects of lead exposure during development on neocortical dendritic and synaptic structure. *Exp. Neurol.* **64**, 487–492.
156. Peuschel, S. M. (1974). Neurological and psychomotor functions in children with an increased lead burden. *Environ. Hlth. Persp.* **7**, 13–16.
157. Pueschel, S. M., Kopito, L. and Schwachman, H. (1972). Children with increased lead burden. *J. Amer. Med. Assn.* **222**, 426–466.
158. Piepho, R. W., Ryan, C. F., and Lacz, J. P. (1976). The effects of chronic lead intoxication and the GABA content of the rat CNS. *Pharmacol.* **18**, 125.
159. Popoff, N., Weinberg, S. and Feigin, I. (1963). Pathologic observations in lead encephalopathy. *Neurol.* **13**, 101–112.
160. Posner, H. S., Damstra, T. and Nriagu, J. O. (1978). Human health effects of lead. In: J. O. Nriagu (Ed.), *the Biogeochemistry of Lead in the Environment.* Elsevier North Holland, pp. 173–223.
161. Press, M. (1977). Lead encephalopathy in neonatal long evans rats: Morphologic studies. *J. Neuropath. Exp. Neurol.* **34**, 169–193.
162. Quarterman, J. and Morrison, J. N. (1975). The effects of dietary calcium and phosphorus on the retention and excretion of lead in rats. *Br. J. Nutr.* **34**, 351–362.
163. Rabinowitz, M., Wetherill, G. and Kopple, J. (1975). Absorption, storage and excretion of lead by normal humans. In: D. D. Hemphill, (Ed.), *Trace Substances in Environmental Health—IX,* University of Missouri, Columbia, pp. 361–376.
164. Rafales, L. S., Bornschein, R. L., Michaelson, I. A., Loch, R. K. and Barker, G. F. (1979). Drug induced activity in lead exposed mice. *Pharmacol. Biochem. Behav.* **10**, 95–104.
165. Ramsay, P. B., Krigman, M. R. and Morell, P. (1980). Developmental studies of the uptake of choline, GABA, and dopamine by crude synaptosomal preparations after in vivo or in vitro lead treatment. *Brain Res.* **187**, 383–402.
166. Repko, J. D. and Corum, C. R. (1979). Critical review and evaluation of the neurological and behavioral sequelae of inorganic lead absorption. *CRC Crit. Rev. Toxicol.* **6**, 135–187.
167. Repko, J. D., Corum, C. R., Jones, P. D. and Garcia, Jr., L. S. (1978). The effects of inorganic lead on behavioral and neurologic function. Final Report. DHEW (NIOSH) Publication No. 78-128.
168. Repko, J. D., Morgan, B. B. and Nicholson, J. (1975). Behavioral effects of occupational exposure to lead. DHEW Public Health Service Publication No. 75-164. Washington.
169. Rice, D. C., and Wiles, R. F. (1979). Neonatal low-level lead exposure in monkeys (Macaca Fasicularis): Effect on two-choice non-spatial form discrimination. *J. Environ. Pathol. Toxicol.* **2**, 1195–1203.
170. Rummo, J. H, Rummo, N. J., Routh, D. K. and Brown, J. F. (1979). Behavorial and neurological effects of symptomatic and

asymptomatic lead exposure in children. *Arch. Environ. Hlth.* **34**, 120–124.

171. Rutter, M. (1980). Raised lead levels and impaired cognitive/ behavorial functioning: A review of the evidence. *Dev. Med. Child Neurol.* (Suppl.) **22**, 1–26.

172. Sachs, H. K., Blanksma, L. A., Murray, E. F. and O'Connell, M. J. (1970). Ambulatory treatment of lead poisoning: Report of 1,155 cases. *Pediat.* **46**, 389–396.

173. Sachs, H. K., Krall, V., McCaughran, D. A., Rozenfeld, I. H., Yongsmith, N., Growe, G., Lazar, B. S., Novar, L., O'Connell, L. and Rayson, B. (1978). I. Q. following treatment of lead poisoning: A patient-sibling comparison. *J. Pediat.* **93**, 428–431.

174. Saito, K. (1973). Electroencephalographic studies on petrol intoxication: Comparison between nonleaded and leaded white petrol. *Br. J. Ind. Med.* **30**, 352–358.

175. Sauer, R. M., Zook, B. C. and Garner, F. M. (1970). Demyelinating encephalopathy associated with lead poisoning in nonhuman primates. *Science* **169**, 1091–1093.

176. Sauerhoff, M. W. and Michaelson, I. A. (1973). Hyperactivity and brain catecholamines in lead exposed developing rats. *Science* **182**, 1022–1024.

177. Schroeder, H. A. and Tipton, I. H. (1968). The human body burden of lead. *Arch. Environ. Hlth.* **17**, 965–978.

178. Seppalainen, A. M., Hernberg, S. and Kock, B. (1979). Relationship between blood Pb levels and nerve conduction velocities. *Neurotoxicol.* **1**, 313–332.

179. Seppalainen, A. M., Tola, S., Hernberg, S. and Kock, B. (1975). Subclinical neuropathy at "safe" levels of lead exposure. *Arch. Environ. Hlth.* **30**, 180–183.

180. Sessa, T., Ferrari, E. and Colucci d'Amato, C. (1965). Velocita di conduzione nervosa nei saturnini. *Folia Med.* **48**, 658–668.

181. Shih, T. M. and Hanin, I. (1977). Lead exposure decreases acetylcholine turnover rate in rat brain areas in vivo. *Fed. Proc.* **36**, 977.

182. Shih, T. M. and Hanin, I. (1978a). Chronic lead exposure in immature animals: Neurochemical correlates. *Life Sci.* **23**, 877–888.

183. Shih, T. M. and Hanin, I. (1978b). Effects of chronic lead exposure on levels of acetylcholine and choline and on acetylcholine turnover rate in rat brain areas in vivo. *Psychopharmacol.* **58**, 263–269.

184. Shih, T. M. Khachaturina, Z. S., Barry, H. and Hanin, I. (1976). Cholinergic mediation of the inhibitory effect of methylphenidate on neuronal activity in the reticular formation. *Neuropharmacol.* **15**, 55–60.

185. Shih, T. M., Khatchaturian, Z. S., Reisler, K. L., Rizk, M. and Hanin, I. (1976). Methylphenidate as a cholinergic agonist: Further observations. *Fed. Proc.* **35**, 307.

186. Silbergeld, E. K. (1977). Interactions of lead and calcium on the synaptosomal uptake of dopamine and choline. *Life Sci.* **20**, 309–318.

187. Silbergeld, E. K. and Chisolm, J. J. (1976). Lead poisoning: Altered urinary catecholamine metabolites as indicators of intoxication in mice and children. *Science* **192**, 153–155.

188. Silbergeld, E. K., Fales, J. T. and Goldberg, A. M. (1974a). The effects of lead on the neuromuscular junction. *Neuropharmacol.* **13**, 795–801.

189. Silbergeld, E. K., Fales, J. T. and Goldberg, A. M. (1974b). Evidence for a junctional effect of lead in neuromuscular function. *Nature* (London) **247**, 49–50.

190. Silbergeld, E. K. and Goldberg, A. M. (1974a) Hyperactivity: A lead induced behavior disorder. *Environ. Hlth. Persp.* **7**, 227–232.

191. Silbergeld, E. K. and Goldberg, A. M. (1974b). Lead induced behavioral dysfunction: An animal model of hyperactivity. *Exp. Neurol.* **42**, 146–157.

192. Silbergeld, E. K. and Goldberg, A. M. (1975). Pharmacological and neurochemical investigations of lead induced hyperactivity. *Neuropharmacol.* **14**, 431–444.

193. Silbergeld, E. K. and Hruska, R. E. (1980) Neurochemical investigations of low level lead exposure. In: H. L. Needleman (Ed.), *Low Level Lead Exposure: The Clinical Implications of Current Research*, Raven, New York, pp. 135–157.

194. Silbergeld, E. K., Miller, L. P., Kennedy, S. and Eng, N. (1979). Lead, GABA and seizures: Effects of subencephalopathic lead exposure on seizure sensitivity and GABAergic function. *Environ. Res.* **19**, 371–382.

195. Simpson, J. A., Seaton, D. A. and Adams, S. F. (1964). Response to treatment with chelating agents of anemia, chronic encephalopathy, and myelopathy due to lead poisoning. *J. Neurol. Neurosurg. Psychiat.* **27**, 536–541.

196. Smith, H., Baehner, R., Curney, T. and Majors, W. (1963). The sequelae of pica with and without lead poisoning. *Amer. J. Dis. Child.* **105**, 609–616.

197. Smith, H. V. (1972). Effects of environmental enrichment on open field activity and Hebb Williams problem solving in rats. *J. Comp. Physiol. Psychol.* **80**, 163–168.

198. Snowden, C. T (1973). Learning deficits in lead injected rats. *Pharmacol. Biochem. Behav.* **1**, 599–603.

199, Sobotka, T. J., Brodie, R. E. and Cook, M. P. (1975). Psychophysiologic effects of early lead exposure. *Toxicol.* **5**, 175–191.

200. Sobotka, T. J. and Cook, M. P. (1974). Postnatal lead acetate exposure in rats. Possible relationship to minimal brain dysfunction. *Amer. J. Mental Defic.* **79**, 5–9.

201. Stumpf, W. E., Sar, M. and Grant, L. D. (1980). Autoradiographic localization of 210Pb and its decay products in rat forebrain. *Neurotoxicol.* **1**, 593–606.

202. Taylor, D., Nathanson, J., Hoffer, B., Olson, L. and Seiger, A. (1978). Lead blockade of norepinephrine induced inhibition of cerebellar Purkinkje neurons. *J. Pharmacol. Exp. Therapeut.* **206**, 371–381.

203. Telfer, J. G. (1947). Use of BAL in lead poisoning. Preliminary report of one case. *J. Amer. Med. Assoc.* **135**, 835–837.
204. Tennekoon, G., Aitchison, C. S., Frangia, J., Price, D. L and Goldberg, A. M. (1979). Chronic lead intoxication: Effects on developing optic nerve. *Ann. Neurol.* **5**, 558–564.
205. Thawley, D. G., Pratt, S. E. and Selby, L. A. (1978). Antagonistic effect of zinc on increased urine delta-aminolevulinic acid excretion in lead intoxicated rats. *Environ. Res.* **15**, 218–226.
206. Thomas, J. A., Dallenbach, F. D. and Thomas, M. (1971). Consideration on the development of experimental lead encephalopathy. *Virchows Arch. Abt. A. Pathol. Anat.* **352**, 61–74.
207. Thomas, J. A., Dallenbach, F. D. and Thomas, M. (1973). The distribution of radioactive lead (210Pb) in the cerebellum of developing rats. *J. Pathol.* **109**, 49–50.
208. Thurston, D., Middelkamp, J. and Mason, E. (1955). The late effects of lead poisoning. *J. Pediat.* **47**, 413–423.
209. Toews, A. D., Kolber, A., Haywood, J., Krigman, M. R. and Morell, P. (1978). Experimental lead encephalopathy in the suckling rat: Concentration of lead in the cellular fractions enriched in brain capillaries. *Brain Res.* **147**, 131–138.
210. Tola, S., Hernberg, S. and Vesanto, R. (1976). Occupational lead exposure in Findland. IV. Final Report. *Scand. J. Work. Environ. Hlth.* **2**, 115–127.
211. Tompsett, S. L. (1939). The influence of certain constituents of the diet upon the absorption of lead from the alimentary tract. *Biochem. J.* **33**, 1237–1240.
212. Tompsett, S. L. and Chalmers, J. N. M. (1939). Studies on lead mobilization. *Br. J. Exp. Pathol.* **20**, 408–417.
213. Valciukas, J. A., Lilas, R., Eisinger, J., Blumberg, W. E., Fischbein, A. and Selikoff, I. J. (1978). Behavioral indicators of lead neurotoxicity: Results of a clinical field survey. *Int. Arch. Occ. Environ. Hlth.* **41**, 217–236.
214. Valciukas, J. A., Lilas, R., Fischbein, A., Selikoff, I. J., Eisinger, J. and Blumberg, W. E. (1978). Central nervous system dysfunction due to lead exposure. *Science* **201**, 465–467.
215. Volkmar, F. R. and Greenough, W. T. (1972). Rearing complexity affects branching of dendrites in the visual cortex of the rat. *Science* **176**, 1445–1447.
216. Waldron, H. A. and Stoffen, D. (1974). *Subclinical Lead Poisoning.* Academic Press, London.
217. Weinreich, K., Stelte, W. and Bitsch, I. (1977). Effects of lead acetate on the spontaneous activity of young rats. *Nutr. Metabol.* **21**, 201–203.
218. Wender, P. H. (1971). *Minimal Brain Dysfunction in Children.* Wiley, New York.
219. West, R. W. and Greenough, W. T. (1972). Effect of environmental complexity on cortical synapses of rats: Preliminary results. *Behav. Biol.,* **7**, 279–284.

220. Will, B. E., Rosenzweig, M. R. and Bennett, E. L. (1976). Effects of differential environments on recovery from neonatal brain lesions, measured by problem solving scores and brain dimensions. *Physiol. Behav.* **16**, 603–611.

221. Will, B. E., Rosenzweig, M. R., Bennett, E. L., Hebert, M. and Morimoto, H. (1977). Relatively brief environmental enrichment aids recovery of learning capacity and alters brain measures after post-weaning brain lesions in rats. *J. Comp. Physiol. Psychol.* **91**, 33–50.

222. Willoughby, R. A., MacDonald, E., McSherry, B. J. and Brown, G. (1972). Lead poisoning and the interaction between lead and zinc poisoning in the foal. *Can. J. Comp. Med.* **36**, 348–359.

223. Wince, L. and Azzaro, A. J. (1978). Neurochemical changes of the central dopamine synapse following chronic lead exposure. *Neurol.* **28**, 382.

224. Windebank, A. J. and Dyck, P. J. (1981). Kinetics of 210Pb entry into the endoneurium. *Brain Res.* **225**, 67–73.

225. Windebank, A. J., McCall, J. T., Hunder, H. G. and Dyck, P. J. (1980). The endoneurial content of lead related to the onset and severity of segmental demyelination. *J. Neuropath. Exp. Neurol.* **39**, 692–699.

226. World Health Organization (1977). *Environ. Health Criteria. 3. Lead.* Geneva.

227. Xintaras, C., Sobecki, M. F. and Ulrich, C. E. (1967). Sleep: Changes in rapid eye movement phase in chronic lead absorption. *Toxicol. Appl. Pharmacol.* **10**, 384.

228. Zenick, H., Rodriquez, W., Ward, J. and Elkington, B. (1979). Deficits in fixed-interval performance following prenatal and postnatal lead exposure. *Dev. Psychobiol.* **12**, 509–514.

229. Zook, B. C., Carpenter, J. L. and Roberts, R. M. (1972). Lead poisoning in dogs; Occurrence, source, clinical pathology and electroencephalography. *Amer. J. Vet. Res.* **33**, 891–902.

Chapter 5

Neurotoxic Effects of Lead

P. McConnell

1. Introduction

Knowledge of the toxicity of lead dates from at least the second century BC, while detailed clinical descriptions of saturnism have been available since the early 1800s (see refs. 6 and 119 for reviews of historical literature). It was not until the beginning of this century, however, that lead was recognized as a particular hazard to the developing infant[41] and the specific syndrome of childhood lead poisoning was described[18].

More recently, improved safety standards in the industrial use of lead and a decline in its domestic applications, have produced a reduction in the incidence of frank lead intoxication in both adults and children. At the same time, though, the dissemination of lead into the general environment has markedly increased.[81,121] The effects of lead exposure in childhood have become, therefore, a focus of increasing attention, with mounting concern that lead levels insufficient to produce overt clinical signs may nevertheless be responsible for more subtle, long-term changes in child development.[28,79,80,121] Hence, despite the present-day rarity of classical clinical lead poisoning, the study of lead toxicity retains direct contemporary relevance.

P. McConnell: Department of Anatomy, Medical School, University of Birmingham, Birmingham, England. Current address: Department of Human Anatomy, University of Oxford, Oxford, England.

2. Clinical Lead Poisoning

The signs and symptoms of frank lead intoxication differ, some-what, between adults and children, although in both cases they are largely nonspecific. The lead poisoned adult commonly presents with pallor, abdominal pain, constipation, and vomiting, with anemia and the frequent presence of a blue "lead line" on the gums.[27] The neurological effects of plumbism, although variable, often take the form of a peripheral neuropathy, generally manifest as a motor disturbance with unilateral or unequally bilateral paralysis of extensor muscle groups in the distal parts of the limbs. The actual distribution of affected muscles appears to be influenced by muscle use, hence wrist drop and weakness of the small muscles of the hand were found to be characteristic symptoms amongst house painters and workers in the lead industry.[85] Occasionally, sensory changes such as paraesthesia, or patches of analgesia and anesthesia may also be present.[27,36,85]

The development of encephalopathy caused by lead, although sometimes occurring in adulthood, is more commonly associated with childhood lead poisoning. In such cases,[24,47] fatigue, pallor, anorexia, and irritability are generally followed by abdominal pain, vomiting, drowsiness, and motor unsteadiness, occasionally complicated by muscular weakness due to a peripheral neuropathy, principally affecting the legs.[82,85,103] If exposure to lead continues, these effects may culminate in the more severe signs of encephalopathy, i.e., convulsions, stupor, coma, and ultimately death.

Survivors of an acute encephalopathic episode of lead poisoning have a poor prognosis, many such children being left with permanent neurological and psychological impairments. The long-term sequelae of clinical lead intoxication (reviewed in refs. 78, 120, and 121) include mental retardation, convulsions, cerebral palsy, optic atrophy, and behavioral and learning disorders, with estimates of the proportion of cases incurring such damage varying from about 25 to over 90%. The effects are seen most frequently following overt encephalopathy, although children showing symptomatic lead poisoning without encephalopathy may also be affected.

3. Neuropathology of Clinical Lead Poisoning

3.1. Peripheral Nervous System

There have been numerous attempts to identify the anatomical changes underlying the neuromuscular dysfunction of clinical lead poisoning. Indeed, the aim of many early studies was simply to establish whether the principal site of action of lead was neuronal or muscular (reviewed in refs. 6 and 52), a debate that continued until comparatively recently,[96] despite accumulating postmortem evidence against the idea of a primary myopathy.[52]

There is now little doubt that lead causes paralysis by its action on nerves, rather than by effects on muscles.[23] In humans, pathological changes of both anterior horn cells[38,49,63] and peripheral nerves[38,45,63] have been described, with the principal lesion being a Wallerian-type axonal degeneration. Segmental demyelination of peripheral nerves has not been reported, although in other species, including the guinea-pig,[40,46] rat,[60,66,76,77,83,102] and cat,[37] segmental demyelination was found to predominate over varying degrees of axonal degeneration.

A mechanism that might account for the variable association between these two degenerative phenomena has recently been proposed by Cavanagh[23] on the basis of the results obtained by Ohnishi et al.[83] in a study of the peroneal and sural nerves of lead-poisoned rats. These latter workers found evidence of random segmental demyelination,[35] the onset of which coincided with the development of endoneurial edema of the affected nerves, but appeared to be independent of signs of axonal disease. Cavanagh, therefore, suggested that lead neuropathy might be similar to that caused by diphtheria toxin, i.e., might involve random metabolic damage to Schwann cells, leading to progressive demyelination and remyelination accompanied by endoneurial swelling. If so, he proposed that axonal degeneration could arise as a result of this edema, since chronically swollen nerves would be susceptible to injury by localized stresses such as minor trauma, pressure at vulnerable sites or heavy muscular work.

It is difficult to understand, however, how the mechanism that Cavanagh suggests could explain the pathology of lead neuropathy in humans[38,45,63] and in rabbits,[117] where axonal degeneration is observed in the apparent absence of segmental demyelination. A more recent proposal,[76] that one of the initial pathological events in lead neuropathy is injury to the blood–nerve barrier, leading to accumulation of edema fluid, seems more readily able to account for these observations. Both axonal degeneration and demyelination might then be secondary to initial vascular damage, demyelination arising from either increased endoneurial pressure[35,66] or a direct toxic effect of the lead-containing edema fluid on Schwann cells.[35,76]

3.2. Central Nervous System

3.2.1. VASCULAR EFFECTS

A comprehensive description of the pathological changes occurring in the human central nervous system (CNS) as a result of clinical lead poisoning was provided by Blackman[19] in 1937, based on observations of 22 children dying of acute lead encephalopathy, and this has been confirmed by subsequent studies.[2,25,84,88,93,107] Lesions, although occurring throughout the brain, are most common in the cerebrum and cerebellum. As is now thought to be the case in the peripheral nervous system (PNS),[76] the primary CNS pathology appears to be vascular. Capillaries are dilated or narrowed, often necrosed and thrombotic, with focal hemorrhages and the accumulation of serous fluid in the perivascular, interstitial, and subpial spaces. Any changes in neurons and glia are generally thought to occur secondary to these effects of lead on small caliber blood vessels.

Because such descriptions of the human pathology can provide only a limited amount of information about the toxic effects of lead, the majority of histopathological studies have concentrated on animal models of the human disease. Only in this way is it possible practically to investigate the vast range of CNS disorders produced by lead, to document the sequence of pathological changes that precede encephalopathy,and to perform meaningful quantitative studies of the effects of lead. The experimental model in most common use is that described by Pentschew and Garrow,[89] in which the addition of a lead compound to the diet of lactating rats produces in the offspring an encephalopathy with histological features similar to those seen in lead-poisoned children. The lead-exposed pups quickly show a retardation in body growth, later developing an abnormally ruffled coat and displaying a broad-based gait[72,74] that eventually gives way to weakness of the hind limbs and finally to frank paralysis, usually between 20 and 30 d postpartum, depending on the dose of lead.

Examination of the CNS at the peak of the illness reveals that, as in the human situation, the pathological changes are widespread, but the cerebellum is particularly susceptible.[1,42,51,55,56,72,1,74,1,42] Again the primary effect of lead appears to be on the developing capillary network, with alterations in the function and morphology of the capillary endothelial cells and basement membrane being detected early in the preparalytic stage of encephalopathy.[1,25,42,58,59,89,94,112,113,118] Studies of capillary-microvessel fractions isolated from the brains of lead poisoned rat pups indicate that lead becomes concentrated in capillary endothelial cells relative to other brain structures[114] and it has been suggested, on the basis of in vitro studies of such fractions, that ensuing abnormalities of calcium homeostasis in these cells may underlie the observed changes in capillary form and function.[43] More recently, comparisons of immature and mature cerebral and cerebellar mitochondrial fractions from lead-exposed animals have shown that lead accumulates within the mitochondrial compartment of immature cells, hence alterations in cellular energy metabolism may be involved in the pathogenesis of lead encephalopathy.[51] Such alterations might occur preferentially within the lead-enriched capillary endothelial cells.

3.2.2. NEURONAL AND GLIAL EFFECTS

Vascular alterations are not the only morphological changes occurring as a result of lead exposure; experimental studies have also revealed abnormalities of neuronal maturation,[55–57,71,91,95,110,112] disturbances of myelination,[55,57,115] and alterations in glial development.[56,110,112,123] Krigman et al.[56] and Krigman and Hogan,[57] in a study of the somatosensory cortex of 30-d-old rats suckled by dams exposed to 4% lead carbonate from delivery, described a general decrease in cortical grey matter, with a relative astyrocytosis. Neuronal numbers were unaffected by the treatment, but their growth and maturation was found to be abnormal. Perikarya were smaller and more densely packed and there was a reduction in the average number of synapses per neuron. The results of this study led the authors to suggest that there might also be a reduction in the extent of the neuronal dendritic trees in the experimental animals and that such alterations of neuronal connectivity might contribute to the latent effects of lead encephalopathy on CNS performance in children. Confirmatory evidence of abnormal dendritic development of neocortical pyramidal cells in lead-poisoned animals has recently been provided by Petit and LeBoutillier.[91]

As with the vascular pathology, the most pronounced neuronal and glial effects of lead poisoning are seen in the cerebellum. Cerebellar weight and total cell number are reduced[72] and the size[71] and morphology of Purkinje cell bodies are abnormal.[110,112] The interneuronal circuitry of the cerebellum is also altered, since the dendritic development of the Purkinje cells, which form the only efferent elements of the cerebellar cortex, is markedly disturbed by lead exposure. Press[95] observed a decrease in the rate of Purkinje cell maturation in lead-poisoned rat pups from about 5 d onwards such that, at 10 d, the leaded cells retained more perisomatic processes than controls and the dendritic arborization was neither as tall nor as extensively branched. The synaptic maturation of the cells was also impaired, with delays in the formation of climbing fiber synapses on primary dendrites and in the establishment of basket cell contacts.

The abnormalities of Purkinje cell dendritic development in lead-poisoned animals were further explored by McConnell and Berry.[71] Qualitative observations of Purkinje cells from 30-d rat pups, suckled by dams given a 4% lead acetate diet, generally revealed a striking difference in dendritic network size relative to controls (Fig. 1). When quantified by the method of network analysis,[13,17] there was found to be a deficit of 35% in network size, total dendritic length being reduced from 7814 ± 310 μm in the control group to 5097 ± 288 μm in the lead-exposed group. This effect was caused partly by a general reduction in segment frequency (Fig. 2) and partly by a reduction in segment length at the periphery of the dendritic network, i.e., in the length of dendritic segments of Strahler order one. These deficits also resulted in a significant decrease in first order pathlength (i.e., the mean length of the paths from the axon hillock to each dendritic terminal[50]). Dendritic density and the frequency of trichotomous branching, on the other hand, were unaffected by the experimental treatment (Table 1), as was also true of the density of dendritic spines.

These abnormalities appear to be the result of metabolic alterations in the Purkinje cells. It has been suggested that dendritic morphology in general is dependent not only upon neuronal metabolism, but also upon the degree of axodendritic interaction during development.[13] In the case of Purkinje cells, the extent of dendritic growth appears to be influenced by the size of the granule cell population[3,4,5,15,20,48,104] since this determines the number of afferent axons (parallel fibers) deposited about the developing dendritic network. However, since the lead-exposed animals showed only a slight, nonsignificant trend towards a reduction in granule cell numbers, it seems probable that the reduction in Purkinje cell size is a result solely of metabolic alterations, either at the level of the dendritic growth cone or the soma. Although the possibility that parallel fiber length is reduced in leaded animals cannot be entirely excluded, if the observed deficit in dendritic segment number had been brought about by a reduction in the density of afferent axons about the developing dendritic terminals, a general increase in the length of spiny segments and a decrease in the frequency of trichotomous nodes would also be expected.[13,15,20] Also, if Purkinje cell metabolism had not been affected by the experimental treatment, pathlengths in these speci-

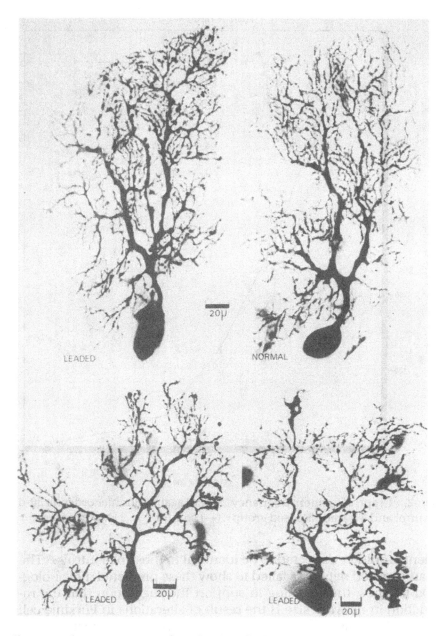

FIG. 1. Photomontages of Purkinje cell dendritic networks from one 30-d control and three 30-d, 4% lead-exposed animals. The majority of lead-exposed Purkinje cell networks are markedly reduced in size relative to controls, although some leaded dendritic fields appear normal.

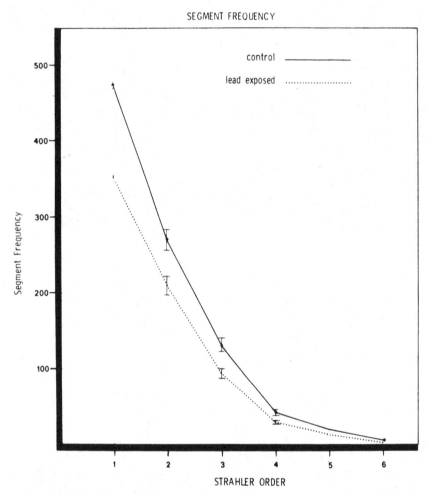

SEGMENT FREQUENCY

FIG. 2.   Graph of segment frequency (±SEM) against Strahler order for 30-d control and 4% lead-exposed groups: (—) control, (. . . .) lead-exposed.

mens would be expected to be identical to those of controls.[20] The lead-exposed networks failed to show these predicted morphological features, thus tending to support the suggestion that the reduction in network size is the result of alterations in Purkinje cell metabolism rather than an alteration in the axonal environment.

   Topological analysis of the branching patterns of lead-exposed Purkinje cell dendritic networks revealed a marked difference from that of control networks.[71] The nature of this difference has been elaborated by reanalyzing the experimental data using the recently developed method of vertex analysis[16] (Fig. 3). This revealed an increase in the degree of collateral branching relative

Table 1
Results of Network Analysis of 30-d Control and 4% Lead-Exposed Dendritic Trees

| | Total number of segments | Total dendritic length, μm | Segment length[a] μm Strahler order | | | | | | First-order path length,[a] μm | Dendritic density, μm/μm² | Percent trichotomous modes |
			1	2	3	4	5	6			
Control	933.6 ±40.74	7814.1 ±310.24	9.54 ±0.50	6.82 ±0.16	6.36 ±0.44	7.91 ±0.24	12.31 ±1.43	19.44 ±1.88	180.50 ±4.07	0.2682 ±0.0117	7.906 ±0.983
4% Leaded	705.8 ±41.05[b]	5096.7 ±288.23[b]	8.41 ±0.17[b]	6.56 ±0.67	6.93 ±0.86	8.95 ±0.98	18.87 ±9.17	28.12 ±3.81	144.53 ±4.05[b]	0.2541 ±0.0084	7.097 ±0.816

[a]Only measurements from 6th-order networks were included in the calculation of these mean values.
[b]Denotes statistically significant difference ($P < 0.05$) between the groups.

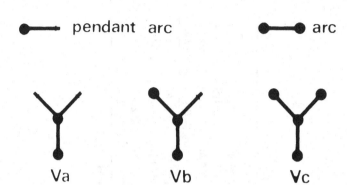

FIG. 3. Vertex classification used in the method of vertex analysis.[16] The method defines the vertices (points interconnected by segments) of a dichotomously branching network according to the number of terminals they connect. A primary vertex (Va) connects two terminal segments and one link segment; a secondary vertex (Vb) connects one terminal segment and two linking segments; a teriary vertex (Vc) connects three linking segments. The frequency of these vertex types within a network varies according to the mode of growth of the network.

to the degree of symmetrical branching, particularly in the proximal portion of the dendritic tree (Fig. 4).

Obviously a number of factors, both extrinsic and intrinsic to the Purkinje cell, could be operating to bring about this increased asymmetry of branching.[16] However, in view of the above evidence suggesting that lead disrupts Purkinje cell metabolism, it might be that the normal functioning of growth cones is disturbed in the lead-exposed animals, resulting in an increase in the extent of nonrandom branching, or the intrusion of a significant degree of non-terminal branching within the developing dendritic network.

3.3. Methodological Criticisms of Experimental Neuropathological Studies of Lead Poisoning.

The major criticism leveled at the Pentschew and Garro[89] model of lead intoxication is that it is confounded by undernutrition of the suckling pups.[106] Many studies have shown that animals lead-poisoned according to this regime rapidly exhibit a reduction in their rate of growth,[1,55,56,58,71,72,74,87,89,94,105,112,113] and it is generally thought that, rather than representing a specific effect of lead on nutritional parameters, this retardation is an artifact of the technique of lead exposure. Administration of lead in the food or water of lactating dams leads to a decline in their food[73,74] and/or fluid consumption[67,73] that, in turn, causes a reduction in the growth of the offspring. Hence, it is argued[23,44,73,74,108] that the histopathological alterations detected in comparisons of such

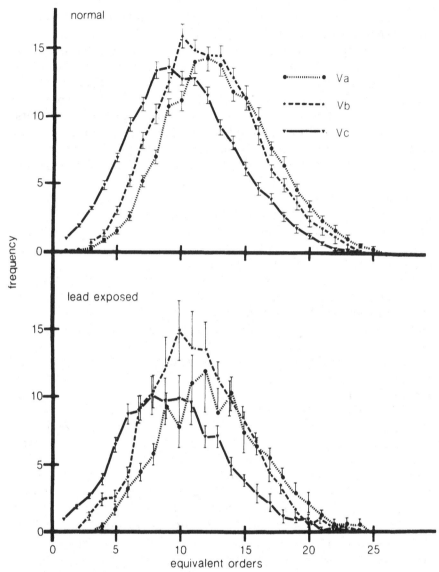

FIG. 4. Distributions of equivalent orders of primary (Va), secondary (Vb), and tertiary (Vc) vertices in control and 4% lead-exposed Purkinje cell dendritic networks. The frequency of secondary vertices relative to primary and tertiary vertices is increased in the lead-exposed animals, particularly over the lower order vertices (i.e., in the proximal portion of the tree), indicating an increase in the degree of collateral branching relative to symmetrical branching.

lead-exposed pups with normally fed controls may, at least in part, represent the pathology of undernutrition rather than that of saturnism.

This criticism is countered to some extent by quantitative comparison of the effects of undernutrition and lead intoxication on the developing brain, since these reveal distinct differences between the two treatments. For example, morphometric analysis of the somatosensory cortex and pyramidal tract from groups of 30-d control, lead-exposed, and undernourished rat pups have demonstrated differential effects of the two experimental treatments upon glial proliferation, synaptic morphology, and the sequence and extent of myelination.[54] Similarly, McConnell and Berry[70] found that in undernourished animals, a reduction in Purkinje cell dendritic network size, comparable in extent to that shown by lead-intoxicated pups,[71] was associated with a 75% deficit in body weight and reductions in cerebellar weight, cerebellar area, granule cell number and Purkinje cell body size. The lead-exposed animals showed a body weight deficit of only 28%. Thus, even if this growth deficit is caused by undernutrition, it is unlikely that such comparatively mild nutritional restriction could account for the observed dendritic changes. The leaded animals showed no change in any of the other cerebellar parmeters known to be affected by severe undernutrition, except in Purkinje cell size, which was increased.

The abnormalities of neuronal growth and connectivity detected in experimental studies of lead-poisoning cannot,then, be attributed to the nutritional complications of the experimental model. Similarly, the altered vascular morphology of leaded animals must represent a direct effect of lead, since such changes, which are widespread in the lead-poisoned human brain (see above), do not occur in nutritionally deprived animals.[23] The Pentschew and Garro[89] model of lead poisoning cannot, therefore, be dismissed as unable to answer questions concerning the specific pathological effects of clinically toxic levels of lead. Whether the model is also appropriate to the study of the currently controversial topic of the possible dangers of low-level lead exposure is a more difficult problem that will be considered below.

4. Subclinical Lead Exposure

There is a growing body of evidence that suggests that the performance of young children in neuropsychological tests may be impaired by exposure to lead levels below those that produce obvious clinical symptoms.[30,79,80] The literature indicates that there is a relationship between such "asymptomatic" or "subclinical" elevation of blood lead levels and a number of CNS effects, including the impairment of fine motor coordination,[21,22,61,97] hyper-activity,[7,21,22,29] learning disability,[90,92] and mild and borderline mental retardation.[10,21,22,30,31] There are, however, several reports in which workers have been unable to demonstrate significant associations between blood lead elevation and various aspects of CNS dysfunction[7,53,61,62,98,99] and many of the studies, both positive and negative, may be criticized on grounds of methodology or logic.[120] Hence, at the present time, there is much debate about the precise extent of the danger posed by "subclinical" lead exposure. It is generally agreed that neuropsychiatric dysfunction may occur in asymptomatic children whose blood lead values are markedly elevated, but there remains fierce controversy over the possible hazards to child development posed by prevailing urban levels of lead.[30,120]

5. Neuropathology of Subclinical Lead Exposure

5.1. Experimental Studies

There have been few attempts to determine the morphological consequences of asymptomatic lead exposure, possibly because of the necessary reliance of such studies upon the use of animal models, the validity of which is open to question. For example, the construction of such models requires the definition of asymptomatic exposure in animals; this, in itself, is a matter for debate.

If it is accepted, for instance, that the reduced growth rate of animals exposed to lead via the Pentschew and Garro[89] method is merely an artifact of the technique of lead administration, it could be argued that such leaded animals, whose growth rate is reduced but who fail to show any other signs of clinical lead poisoning, are actually asymptomatic with respect to lead. If this is so, any neuropathological consequences of low-level lead exposure should be revealed by detailed morphometric comparisons[54] between lead-exposed animals and pair-fed controls. On the other hand, if the growth deficit is viewed as a specific, lead-related effect,[65,105] it becomes necessary to administer lead at such a level and in such a way as to avoid this effect.

Both of the above philosophies have been employed in experiments to assess the effects of subclinical lead exposure on CNS morphology, with essentially similar results. Prompted by the observation that Purkinje cell dendritic morphology is markedly abnormal in experimentally lead-poisoned rat pups,[71] McConnell[69] studied the dendritic morphology of Purkinje cells from a small number of animals that had been exposed to lower levels of lead acetate (2 and 0.5%) via the maternal drinking water. As expected from studies by previous workers,[67] the growth of these animals was reduced (Fig. 5) relative to that of normally fed controls raised in litters of comparable size (six pups/litter). At neither of the levels of lead exposure, however, did animals exhibit any signs of overt clinical intoxication, nor were there any indications of gross CNS pathology, despite substantially elevated blood lead levels at 30 d (2% leaded group, 222 μg lead/100 mL blood; 0.5% leaded group, 109 μg/100 mL; controls, 5 μg/100 mL). The animals were therefore deemed asymptomatic with respect to lead.

Quantitative dendritic analysis of Purkinje cells from these leaded animals, however, indicated that at least the higher of the two apparently subclinical levels of lead exposure was associated with abnormalities of Purkinje cell development (Table 2). Both groups, in fact, showed deficits in mean total dendritic length and segment frequency relative to normally fed controls, but the number of cells in the 0.5% lead group was too small to enable the statistical significance of these deficits to be determined. Unfortunately, though, the failure of this limited study to include pair-fed control groups, makes it impossible to assess any contribution of undernutrition to the dendritic abnormalities of the leaded pups. Thus, while it seems unlikely, in the 2% group at least, that undernutrition is the sole cause of the aberrant dendritic development (a comparable dendritic length deficit in undernourished animals[70] was found to be associated with a much greater reduction in body weight), the study provides only tentative evidence of neuropathological change following early, low-level lead exposure.

Supporting evidence of neuronal abnormalities comes from a study by Louis-Ferdinand et al.,[65] taking as a model of subclinical lead exposure rat pups given lead acetate solution by daily injection at a "dose which did not impair weight gain" (7.5 mg/kg ip from 0–10 d postpartum). Thirty-day pups were found to display localized deficits in the thickness and cellularity of the hippocampal pyramidal layer and the granular layer of the dentate gyrus, together with a reduction in neocortical thickness. The authors suggested that this latter deficit might be attributable to defects in the myelination of fibers in the cortical neuropil, a hypothesis that receives support from a study by Tennekoon et al.[111] of pups raised in litters of reduced size by dams given 0.5% lead acetate solution to drink. These animals, whose growth was not retarded by the experimental procedure, and who displayed no

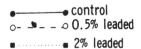

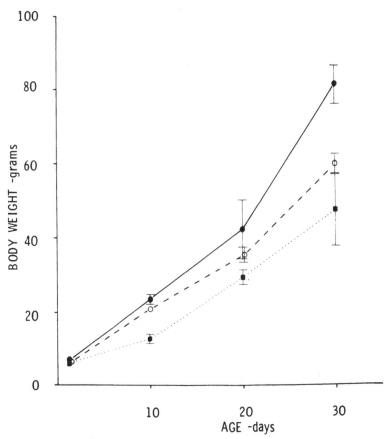

Fɪɢ. 5.   Graph of increase in body weight with age in normally fed control, 0.5% lead-exposed, and 2% lead-exposed rat pups.

signs of overt lead poisoning, showed hypomyelination of the optic nerve, together with at reduction in the diameter of optic axons.

5.2. Extrapolation to the Human Situation

The few available studies of the morphological consequences of low-level lead exposure indicate that young rats, exposed to lead levels insufficient to produce overt signs of clinical toxicity, show

Table 2
Results of Network Analysis of 30-d Control, 2% Lead-Exposed, and 0.5% Lead-Exposed Dendritic Networks

| | Total number of segments | Total dendritic length, μm | Segment length,[a] μm | | | | | | First-order path length,[a] μm | Dendritic density, μm/μm² | Percent trichotomous nodes |
			1	2	3	4	5	6			
Control	933.6 ±40.74	7814.1 ±310.24	9.54 ±0.50	6.82 ±0.16	6.36 ±0.44	7.91 ±0.24	12.31 ±1.43	19.44 ±1.88	180.50 ±4.07	0.2682 ±0.0117	7.906 ±0.983
2% Leaded	711.00 ±23.62[b]	4624.6 ±57.11[b]	7.79 ±0.25[b]	5.24 ±0.16[b]	5.6 ±0.35	6.66 ±0.69[b]	10.01 ±1.11	18.34 ±2.79	140.51 ±6.60	0.2841 ±0.0218	8.5032 ±1.0387
0.5% Leaded	882.00 ±85.95	6487.86 ±712.25	8.47 ±0.80	6.31 ±0.67	5.93 ±0.31	6.74 ±0.42	9.57 ±1.93	21.29 ±1.14	150.00 ±8.19	0.2732 ±0.0184	5.2146 ±0.5607

[a]Only measurements from 6th-order networks were included in the calculation of these mean values.

[b]Denotes statistically significant difference ($P < 0.05$) when compared with controls. The number of 6th-order trees in the 0.5% leaded group was too low to allow statistical comparison with the control group. Comparison of many of the parameters measured by network analysis is valid only between networks attaining the same maximum Strahler order; the majority of control cells were 6th order.

abnormalities of neuronal growth and connectivity. Although the functional consequences of these findings are presently unknown, any suggestion that similar changes in neuronal development might occur in subclinically lead-exposed children could obviously have far-reaching social, political, and economic implications, in view of the recent recognition that large numbers of children in urban environments have elevated blood and/or body lead levels.[24,64,73,97,121] It is therefore necessary to examine particularly critically the relevance of the various experimental models to the human situation.

Unlike animal models of clinical lead poisoning, the validity of experimental models of low-level lead exposure obviously cannot be confirmed by direct comparison with human material. The relevance of such models to low-level lead exposure in humans must therefore be assessed by considering their comparability in terms of the severity, duration, and timing of the insult.

Interspecies equation of the timing of exogenous influences on CNS development is fraught with difficulties, since the events involved in brain maturation occur at different times in relation to birth in different species.[32] It is necessary, therefore, to consider the timing of the insult in relation to the stages of brain maturation, rather than its relation to birth.

Within the human population, the group thought to be most at risk of exposure to high levels of lead is that comprising infants and preschool children. There is a common tendency among such young children to explore the environment with the mouth, a practice that often involves the ingestion of nonfood substances. This activity, know as "pica", develops when the child becomes mobile and shows a decline with increasing age, i.e., the incidence of pica is negligible before 6–7 months postpartum[26,39] and the habit persists in only 10% of children at 5 yr.[8] It is between these two ages that children, particularly in urban communities, may be exposed to large amounts of lead from the ingestion of lead-containing dust and flakes of plaster and paint.[24,121]

The degree of brain maturation in the newborn rat is comparable to that achieved in the mid-term human fetus.[33] The 7-month-old human infant is probably equivalent, at least in terms of maturation of neuronal connectivity, to a 21-d rat pup, with the major period of dendritic elaboration virtually complete in both the cerebral[68,86,100,101,109,116] and cerebellar cortices.[14,122] Hence, the experimental models of low-level lead exposure used to date, in which rat pups were maintained on a leaded diet from birth until 10,[65] 18,[111] or 30 d postpartum,[69] have relatively little in common with the human situation outlined above.

The experiments equate more closely with the situation described by Beattie et al.[10] and Moore et al.[75] who found evidence of high pre- and postnatal lead exposure in children born to mothers living, throughout pregnancy and the first year of their infant's life, in homes with lead-contaminated water. However, even with respect to this less common form of human lead exposure, the experimental models are imperfect. Placental transfer of lead to the human fetus is known to begin as early as the 12th–14th week of gestation.[9] Hence, in the circumstances outlined above, the fetus may be exposed to elevated lead levels while the major period of neurogenesis is in progress, as well as during the later brain growth spurt.[12,34] In the animal models, lead exposure does not begin until neuronal multiplication is virtually complete in the majority of brain regions.[5]

Nevertheless, as the period of lead exposure begins relatively earlier in infants exposed to lead via the water supply, and is of relatively longer duration than in the animal models, it seems likely that the experimental findings of altered neuronal growth and connectivity could have some parallel in this particular human situation.

Even such tentative extrapolation between animal and human lead exposure may be invalid, however, since further complications arise when comparing the severity of lead exposure in the two situations. Although experimentally induced neurological abnormalities were present in animals that failed to show clinical symptoms of lead toxicity, the soft-tissue levels of lead to which these animals were exposed were probably well beyond the range of common human experience, at least if the relationships between blood and soft-tissue lead in humans are similar to those in the rat. For example, the abnormalities of Purkinje cell dendritic development seen in 30-d rat pups that failed to show any physical signs of lead toxicity[69] were nevertheless associated with a blood lead level of 222 ± 22 µg/100 mL. In the human infants, blood lead levels greater than 80 µg/100 mL are generally associated with severe symptoms of encephalopathy,[24] and children with signs and symptoms of clinical poisoning have been reported with blood lead values below this level.[11] Thus, there appears to be a vast difference in the degree of lead exposure that can be tolerated in these two species, and it is difficult to know to what extent observations on asymptomatic, leaded rats are of relevance to the subclinically lead-exposed child.

6. Conclusion

The inadequacies of existing animal models of low-level lead exposure clearly preclude any firm conclusions concerning the morphological consequences of such insult to the developing human
brain. Nevertheless, in view of the widespread distribution of lead
in the contemporary urban environment,[81,121] even the currently
available, extremely tentative evidence that subclinical levels of
lead may impair neuronal growth and connectivity in the developing brain, serves to emphasize the urgent need for further investigation. Possibly direct studies of the connectivity and brain lead
content of child accident victims should be considered if a more
satisfactory animal model of subclinical lead exposure cannot be
found.

Acknowledgments

I am grateful to Professor Martin Berry for his comments, and to
Ms. Gill Taylor for typing the manuscript.

References

1. Ahrens, F.A. and Vistica, D.T. (1977) Microvascular effects of lead in
   the neonatal rat I. Histochemical and light microscopic studies,
   Exp. Mol. Path. **26**, 129–138.
2. Akelaitis, A.M. (1941) Lead encephalopathy in children and adults: a
   clinicopathological study, *J. Nerv. Ment. Dis.* **93**, 313–332.
3. Altman, J. and Anderson, W.J. (1971) Irradiation of the cerebellum in
   infant rats with low level X-ray: histological and cytological effects in
   infancy and adulthood, *Exp. Neurol.* **30**, 492–509.
4. Altman, J. and Anderson, W.J. (1972) Experimental reorganisation of
   the cerebellar cortex I. Morphological effects of elimination of all
   microneurons with prolonged X-irradiation started at birth, *J.
   Comp. Neurol.* **146**, 355–406.
5. Altman, J. and Das, G.D. (1966) Autroradiographic and histological
   studies of postnatal neurogenesis I. A longitudinal investigation of
   the kinetics, migration and transformation of cells incorporating
   tritiated thymidine in neonate rats, with special reference to postnatal neurogenesis in some brain regions, *J. Comp. Neurol.* **126**,
   337–390.
6. Aub, J.C., Fairhall, L.T., Minot, A.S. and Reznikoff, P. (1925) Lead
   poisoning, *Medicine* **4**, 1–250.

7. Baloh, R., Sturm, R., Green, B. and Gleser, G. (1975) Neuropsychological effects of chronic, asymptomatic increased lead absorption. A controlled study, *Arch. Neurol.* **32**, 326–330.

8. Barltrop, D. (1966) The prevalence of pica, *Am. J. Dis. Child.* **112**, 116–123.

9. Barltrop, D. (1969) Transfer of lead to the human fetus. In: D. Barltrop and W.L. Barland (Eds) *Mineral Metabolism in Pediatrics*, Davis, Philadelphia, pp. 135–151.

10. Beattie, A.D., Moore, M.R., Goldberg, A., Finlayson, M.J.W., Graham, J.F., Mackie, E.M., Main, J.C., McLaren, D.A., Murdoch, R.M. and Steward, G.T. (1975) Role of chronic low-level lead exposure in the aetiology of mental retardation, *Lancet* I, 589–592.

11. Berman, E. (1966) The biochemistry of lead: review of the body distribution and methods of lead determination, *Clin. Pediat.* **5**, 287–291.

12. Berry, M. (1981) The development of the human nervous system. In: J.W.T. Dickerson (Ed) *Brain and Behaviour*, Blackie, London, pp. 6–46.

13. Berry, M. and Bradley, P. (1976) The application of network analysis to the study of branching patterns of large dendritic fields, *Brain Res.* **109**, 111–132.

14. Berry, M. and Bradley, P. (1976) The growth of dendritic trees of Purkinje cells in the cerebellum of the rat, *Brain Res.* **112**, 1–35.

15. Berry, M. and Bradley, P. (1976) The growth of the dendritic trees of Purkinje cells in irradiated agranular cerebellar cortex, *Brain Res.* **116**, 361–387.

16. Berry, M. and Flinn, R. (1982) Vertex analysis of neural networks, *J. Comp. Neurol.* In press.

17. Berry, M., Hollingworth, T., Anderson, E.M. and Flinn, R.M. (1976) Application of network analysis to the study of the branching patterns of dendritic fields. In: G.W. Kreutzberg (Ed) *Advances in Neurology*, Vol. 12, *Physiology and Pathology of Dendrites*, Raven Press, New York, pp. 217–245.

18. Blackfan, K.D. (1917) Lead poisoning in children with especial reference to lead as a cause of convulsions, *Am. J. Med. Sci.* **153**, 877–887.

19. Blackman, S.S. (1937) The lesion of lead encephalitis in children, *Bull. John Hopkins Hosp.* **61**, 1–62.

20. Bradley, P. and Berry, M. (1978) Quantitative effects of methylazoxymethanol acetate on Purkinje cell dendritic growth, *Brain Res.* **143**, 499–511.

21. de la Burdé, B. and Choate, M.S. (1972) Does asymptomatic lead exposure in children have latent sequelae? *J. Pediat.* **81**, 1088–1091.

22. de la Burdé, B. and Choate, M.S. (1975) Early asymptomatic lead exposure and development at school age, *J. Pediat.* **87**, 638–642.

23. Cavanagh, J.B. (1979) Metallic toxicity and the nervous system. In: W. Thomas Smith and J.B. Cavanagh (Eds) *Recent Advances in Neuropathology*, Churchill Livingstone, Edinburgh, pp. 247–275.

24. Chisolm, J.J. (1971) Lead poisoning, *Sci. Amer.* **224**, 2, 15–23.
25. Clasen, R.A., Hartmann, J.F., Starr, A.J., Coogan, P.S., Pandolfi, S., Laing, I., Becker, R. and Hass, G.M. (1974) Electron microscopic and chemical studies of the vascular changes and edema of lead encephalopathy, *Am. J. Pathol.* **74**, 215–240.
26. Cooper, M. (1957) *Pica* Thomas, Springfield, Illinois.
27. Dagg, J., Goldberg, A., Lochhead, A. and Smith, J. (1965) The relationship of lead poisoning to acute intermittent porphyria, *Quart. J. Med.* **134**, 163–175.
28. David, O.J., Clark, J. and Hoffman, S. (1979) Childhood lead poisoning: a re-evaluation, *Arch. Environ. Health* **34**, 106–111.
29. David, O., Clark, J. and Voeller, K. (1972) Lead and hyperactivity, *Lancet* **II**, 1376–1379.
30. David, O.J., Hoffman, S. and Kagey, B. (1978) Subclinical lead effects, paper read at Conservation Society Symposium: Lead pollution —health effects, London.
31. David, O., Hoffman, S., McGann, B., Sverd, J. and Clark, J. (1976) Low lead levels and mental retardation, *Lancet* **II**, 1376–1379.
32. Dobbing, J. (1968) Effects of experimental undernutrition on the development of the nervous system. In: N.S. Scrimshaw and J.E. Gordon (Eds) *Malnutrition, Learning and Behavior*, MIT Press, Cambridge, Mass. pp. 181–202.
33. Dobbing, J. (1973) The developing brain: a plea for more critical interspecies extrapolation, *Nutr. Rep. Int.* **7**, 401–406.
34. Dobbing, J. and Sands, J. (1973) Quantitative growth and development of human brain, *Arch. Dis. Child.* **48**, 757–767.
35. Dyck, P.J., O'Brien, P.C. and Ohnishi, A. (1977) Lead neuropathy: 2. Random distribution of segmental demyelination among old internodes of myelinated fibres, *J. Neuropath. Exp. Neurol.* **36**, 570–575.
36. Feldman, R.G., Hayes, M.K., Younes, R. and Aldrich, F.D. (1977) Lead neuropathy in adults and children, *Arch. Neurol.* **34**, 481–488.
37. Ferraro, A. and Hernandez, R. (1932) Lead poisoning. A histopathological study of the nervous system of cats and monkeys in the acute and subacute stages, *Psychiat. Quart.* **6**, 131–146, 319–350.
38. Fischer, E.D. (1892) Lead poisoning with special reference to the spinal cord and peripheral nerve lesions, *Am. J. Med. Sci.* **104**, 51–54.
39. Freeman, R. (1970) Chronic lead poisoning in children: a review of 90 children diagnosed in Sydney, 1948–1967. *Med. J. Aust.* **I**, 640–647.
40. Fullerton, P.M. (1966) Chronic peripheral neuropathy produced by lead poisoning in guinea-pigs, *J. Neuropath. Exp. Neurol.* **25**, 214–236.
41. Gibson, J.L. (1917) The diagnosis, prophylaxis and treatment of plumbic ocular neuritis among Queensland children, *Med. J. Aust.* **2**, 201–204.
42. Goldstein, G.W., Asbury, A.K. and Diamond, I. (1974) Pathogenesis of lead encephalopathy, *Arch. Neurol.* **31**, 382–389.

43. Goldstein, G.W., Wolinsky, J.S. and Csejtey, J. (1977) Isolated brain capillaries: a model for the study of lead encephalopathy, *Ann. Neurol.* **1**, 235–239.
44. Golter, M. and Michaelson, I. (1975) Growth, behaviour and brain catecholamines in lead exposed neonatal rats: a reappraisal, *Science* **187**, 359–361.
45. Gombault, M. (1873) Contribution à l'histoire anatomique de l'atrophie musculaine saturnine. *Arch. Physiol.* **5**, 592–597.
46. Gombault, M. (1880) Contribution à l'étude anatomique de la névrite parenchymateuse subaigué et chronique - névrite segmentaire péri-axile, *Arch. Neurol.* **1**, 11–38.
47. Graef, J.W. (1975) The prevention of lead poisoning. In: A. Milunsky (Ed), *The Prevention of Genetic Disease and mental retardation*, Saunders, Philadelphia, pp.354–368.
48. Herndon, R.M. and Oster-Granit, M.L. (1975) Effect of granule cell destruction on development and maintenance of the Purkinje cell dendrite. In: G.W. Kreutzberg (Ed) *Advances in Neurology*, Vol. 12, *Physiology and Pathology of Dendrites*, Raven Press, New York, pp. 361–371.
49. Herter, C.A. (1895) Report of a case of lead paralysis, with special reference to cytological changes in the nervous system and the distribution of lead, *NY Med. J.* **61**, 665–667.
50. Hollingworth, T. and Berry, M. (1975) Network analysis of dendritic fields of pyramidal cells in neocortex and Purkinje cells in the cerebellum of the rat, *Phil. Trans. Roy. Soc. (Lond.)* B **270**, 227–264.
51. Holtzman, D., Herman, M.M., Shen Hsu, J. and Mortell, P. (1980) The pathogenesis of lead encephalopathy, *Virchows Arch. A Path. Anat. Histol.* **387**, 147–164.
52. Hyslop, G.H. and Kraus, W.M. (1923) The pathology of motor paralysis by lead, *Arch. Neurol Psychiat.* **10**, 444–455.
53. Kotok, D. (1972) Development of children with elevated blood lead levels: a controlled study, *J. Pediat.* **80**, 57–61.
54. Krigman, M.R. (1978) Neuropathology of heavy metal intoxication, *Env. Health Perspect.* **26**, 117–120.
55. Krigman, M.R., Druse, M.J., Traylor, T.D., Wilson, M.H., Newell, L.R. and Hogan, E.L. (1974) Lead encephalopathy in the developing rat: effect upon myelination, *J. Neuropath. Exp. Neurol.* **33**, 58–73.
56. Krigman, M.R., Druse, M.J., Traylor, T.D., Wilson, M.H., Newell, L.R. and Hogan, E.L. (1974) Lead encephalopathy in the developing rat: effect on cortical ontogenesis, *J. Neuropath. Exp. Neurol.* **33**, 671–686.
57. Krigman, M.R. and Hogan, E.L. (1974) Effect of lead intoxication on the postnatal growth of the rat nervous system, *Env. Health Perspect.***7**, 187–199.
58. Lampert, P. (1966) Lead encephalopathy, *Bull. Pathol.* **7**, 222–223.
59. Lampert, P., Garro, F. and Pentschew, A. (1967) Lead encephalopathy in suckling rats - an electron microscopic study. In: I. Klatzo and

F. Seitelberg (Eds), *Brain Edema*, Springer-Verlag, New York, pp. 207–222.

60. Lampert, P. and Schochet, S.S. (1968) Demyelination and remyelination in lead neuropathy, *J. Neuropath. Exp. Neurol.* **27**, 527–545.

61. Landrigan, P.J., Whitworth, R.H., Baloh, R.W., Staohling, N.W., Barthel, W.F. and Rosenblum, B.F. (1975) Neuropsychological dysfunction in children with chronic low-level lead absorption, *Lancet* **I**, 708–712.

62. Lansdown, R.G., Shepherd, J., Clayton, B.E., Delves, H.T., Graham, P.J. and Turner, W.C. (1974) Blood-lead levels, behaviour and intelligence. A population study, *Lancet* **I**, 538–541.

63. Laslett, E.E.. and Warrington, W.B. (1898) The morbid anatomy of a case of lead paralysis. Condition of the nerves, muscles, muscle spindles and spinal cord, *Brain* **21**, 224–231.

64. Lin-Fu, J.S. (1972) Undue absorption of lead among children—a new look at an old problem, *N. Eng. J. Med.* **286**, 702-710.

65. Louis-Ferdinand, R.T., Brown, D.R., Fiddler, S.F., Daughtrey, W.C. and Klein, A.W. (1978) Morphometric and enzymatic effects of neonatal lead exposure in the rat brain. *Toxicol. Appl. Pharmacol.* **43**, 351–360.

66. Low, P.A. and Dyck, P.J. (1977) Increased endoneurial fluid pressure in experimental lead neuropathy, *Nature* **269**, 427–428.

67. Marker, H.S., Lehrer, G.M. and Silides, D.J. (1975) The effects of lead on mouse brain development, *Environ. Res.* **10**, 76–91.

68. Marin-Padilla, M. (1970) Perinatal and early postnatal ontogenesis of the human motor cortex; A Golgi study. I. The sequential development of the cortical layers, *Brain Res.* **23**, 167–183.

69. McConnell, P. (1979) The effects of undernutrition and lead exposure on the development of the cerebellum, PhD. Thesis, University of Birmingham.

70. McConnell, P. and Berry, M. (1978) Effects of undernutrition on Purkinje cell dendritic growth in the rat, *J. Comp. Neurol.* **177**, 159–172.

71. McConnell, P. and Berry, M. (1979) The effects of postnatal lead exposure on Purkinje cell dendritic development in the rat, *Neuropath. Appl. Neurobiol.* **5**, 115–132.

72. Michaelson, I.A. (1973) Effects of inorganic lead on RNA, DNA and protein content of the developing neonatal rat brain, *Toxicol. Appl. Pharmacol.* **26**, 539–548.

73. Michaelson, I.A. (1980) An appraisal of rodent studies on the behavioural toxicity of lead. The role of nutritional status. In: R.L. Singhal and J.A. Thomas (Eds) *Lead Toxicity*, Urban and Schwarzenberg, Baltimore, pp. 301–365.

74. Michaelson, I.A. and Sauerhoff, M.W. (1974) Animal models of human disease: severe and mild lead encephalopathy in the neonatal rat, *Env. Health Perspect.* **7**, 201–225.

75. Moore, M.R., Meredith, P.A. and Goldberg, A. (1977) A retrospective analysis of blood-lead in mentally retarded children, *Lancet* I, 717–719.

76. Myers, R. R., Powell, H.C., Shapiro, H.M., Costello, M.L. and Lampert, P.W. (1980) Changes in endoneurial fluid pressure, permeability and peripheral nerve ultrastructure in experimental lead neuropathy, *Ann. Neurol.* **8**, 392–401.

77. Nagatoshi, K. (1979) Experimental chronic lead poisoning, *Folia Psychiat. Neurol. Jpn.* **33**, 123–131.

78. Needleman, H.L. (1973) Lead poisoning in children: neurological implications of widespread subclinical intoxication, *Sem. Psychiatry* **5**, 47–54.

79. Needleman, H.L. (1980) Human lead exposure: difficulties and strategies in the assessment of neuropsychological impact. In: R.L. Singhal and J.A. Thomas (Eds), *Lead Toxicity*, Urban and Schwarzenberg, Baltimore, pp. 1–17.

80. Needleman, H.L., Gunnoe, C., Leviton, A., Reed, R., Perseie, H., Maher, C. and Barrett, P. (1979) Deficits in psychologic and classroom performance of children with elevated dentine lead levels, *N. Engl. J. Med.* **300**, 689–695.

81. Nriagu, J.O. (1980) Lead in the atmosphere and its effect on lead in humans. In: R.L. Singhal and J.A. Thomas (Eds.) *Lead Toxicity*, Urban and Schwarzenberg, Baltimore, pp. 483–503.

82. Nye, L.J.J. (1933) Chronic nephritis and lead poisoning, Angus and Robertson, Sydney, pp. 48–66.

83. Ohnishi, A., Schilling, K., Brimijoin, W.S., Lambert, E.H., Fairbanks, W.F. and Dyck, P.J. (1977) Lead neuropathy 1: morphometry, nerve conduction and choline acetyltransferase transport: new finding of endoneurial edema associated with segmental demyelination, *J. Neuropath. Exp. Neurol.* **36**, 499–518,

84. Okazaki, H., Aronson, S.M., Dimaio, D.J. and Olvera, J.E. (1963) Acute lead encephalopathy of childhood, *Trans. Am. Neurol. Assoc.* **88**, 248–250.

85. Oliver, T. (1914) *Lead Poisoning*. Lewis, London.

86. Parnavelas, J.G., and Uylings, H.B.M. (1980) The growth of non-pyramidal neurons in the visual cortex of the rat: a morphometric study. *Brain Res.* **193**, 373–382.

87. Patel, A.J., Michaelson, I.A., Cremer, J.E. and Balazs, R. (1974) The metabolism of (^{14}C) glucose by the brains of suckling rats intoxicated with inorganic lead, *J. Neurochem.* **22**, 581–590.

88. Pentschew, A. (1965) Morphology and morphogenesis of lead encephalopathy, *Acta Neuropath.* **5**, 133–160.

89. Pentschew, A. and Garro, F. (1966) Lead encephalomyelopathy of the suckling rat and its implications on the porphyrinopathic nervous diseases, *Acta Neuropath.* **6**, 266–278.

90. Perino, J. and Ernhart, C.B. (1974) The relation of subclinical lead level to cognitive and sensorimotor impairment in black preschoolers, *J. Learning Disab.* **7**, 616–620.

91. Petit, T.L. and LeBoutillier, J.C. (1979) Effects of lead exposure during development on neocortical dendritic and synaptic structure, *Exp. Neurol.* **64**, 482–492.

92. Pihl, R.O. and Parkes, M. (1977) Hair element content in learning disabled children, *Science* **198**, 204–206.

93. Popoff, N., Weinberg, S. and Feigin, I. (1963) Pathologic observations in lead encephalopathy, *Neurology* **13**, 101–112.

94. Press, M.F. (1977) Lead encephalopathy in neonatal Long-Evans rats: morphologic studies, *J. Neuropathol. Exp. Neurol.* **36**, 169–193.

95. Press, M.F. (1977) Neuronal development in the cerebellum of lead poisoned neonatal rats, *Acta Neuropathol.* **40**, 259–268.

96. Prieskel, D. (1958) Chronic lead poisoning: myopathy or neuritis? *Ann. Phys. Med.* **4**, 293–296.

97. Pueschel, S.M., Kopito, L. and Schwachman, H. (1972) Children with an increased lead burden. A screening and follow up study. *J. Amer. Med. Assn.* **222**, 462–446.

98. Ratcliffe, J.M. (1977) Developmental and behavioural functions in young children with elevated blood lead levels, *Br. J. Prev. Soc. Med.* **31**, 258–264.

99. Rummo, J.H., Rummo, N.J., Routh, D.K. and Brown, J.F. (1979) Behavioural and neurological effects of symptomatic and asymptomatic lead exposure in children, *Arch. Environ. Hlth.* **34**, 120–124.

100. Schadé, J.P., van Backer, H. and Colon, E. (1964) Quantitative analysis of neuronal parameters in the maturing cerebral cortex, *Prog. Brain Res.* **4**, 150–175.

101. Schadé, J.P. and van Groenigen, W.B. (1961) Structural organisation of the human cerebral cortex I. Maturation of the middle frontal gyrus, *Acta Anat.* **47**, 74–111.

102. Schlaepfer, W.W. (1969) Experimental lead neuropathy: a disease of the supporting cells in the peripheral nervous system, *J. Neuropathol. Exp. Neurol.* **28**, 401–418.

103. Seto, D.S.Y. and Freeman, J.M. (1964) Lead neuropathy in childhood, *Amer. J. Dis. Child.* **107**, 337–342.

104. Shofer, R.J., Pappas, G.D. and Purpura, D.P. (1964) Radiation induced changes in morphological and physiological properties of immature cerebellar cortex. In: J.J. Haley and R.S. Snider (Eds) *Response of the Nervous System to Ionizing Radiation*, Little, Brown, New York, pp. 376–508.

105. Silbergeld, E.K. and Goldberg, A.M. (1974) Hyperactivity: a lead induced behaviour disorder, *Env. Health Perspect.* **7**, 227–232.

106. Silbergeld, E.K. and Goldberg, A.M. (1980) Problems in experimental studies of lead poisoning. In: R.L. Singhal and J.A. Thomas (Eds.) *Lead Toxicity*, Urban and Schwarzenberg, Baltimore, pp. 19–41.

107. Smith, J.F., Mclauren, R.L., Nichols, J.B. and Asbury, A. (1960) Studies in cerebral oedema and cerebral swelling I. The changes in lead encephalopathy in children compared with those in alkyl tin poisoning in animals, *Brain* **83**, 411–424.

108. Sobotka, T.J. and Cook, M.P. (1974) Postnatal lead acetate exposure in rats: possible relationship to minimal brain dysfunction, *Am. J. Ment. Defic.* **79**, 5–9.

109. Takashima, S., Chan, F., Becker, L.E. and Armstrong, D.L. (1980) Morphology of the developing visual cortex of the human infant. A quantitative and qualitative Golgi study, *J. Neuropathol. Exp. Neur.* **39**, 487–501.

110. Takeichi, M. and Noda, Y. (1974) Electron microscopy of experimental lead encephalopathy—consideration on the development mechanism of brain lesions, *Folia Psych. Neurol. Jap.* **28**, 217–232.

111. Tennekoon, G., Aitchison, C.S., Frangia, J., Price, D.L. and Goldberg, A.M. (1979) Chronic lead intoxication: effects on developing optic nerve, *Ann. Neurol.* **5**, 558–564.

112. Thomas, J.A., Dallenbach, F.D. and Thomas, M. (1971) Considerations on the development of experimental lead encephalopathy, *Virch. Arch. A. Path. Anat. Histol.* **352**, 61–74.

113. Thomas, J.A. and Thomas, M. (1974) The pathogenesis of lead encephalopathy, *Ind. J. Med. Res.* **62**, 36–41.

114. Toews, A.D., Kolber, A., Hayward, J., Krigman, M.R. and Morell, P. (1978) Experimental lead encephalopathy in the suckling rat: concentration of lead in cellular fractions enriched in brain capillaries, *Brain Res.* **147**, 131–138.

115. Toews, A.D., Krigman, M.R., Thomas, D.J. and Morell, P. (1980) Effect of inorganic lead exposure on myelination in the rat, *Neurochem. Res.* **5**, 605–616.

116. Uylings, H.B.M. and Parnavelas, J.G. (1981) Growth and plasticity of cortical dendrites. In: O. Fehér and F. Joó (Eds) *Cellular Analogues of Conditioning and Neural Plasticity*, Pergamon Press.

117. de Villaverde, J.M. (1930) Sur l'avenir des parties constitutives de la fibre neuveuse dans l'intoxication experimentale par le plomb, *Trav. Inst. Cajal Invest. Biol.* **26**, 163–187.

118. Vistica, D.T. and Ahrens, F.A. (1977) Microvascular effects of lead in the neonatal rat II. An ultrastructural study, *Exp. Mol. Pathol.* **26**, 139–151.

119. Waldron, H.A. (1973) Lead poisoning in the ancient world, *Med. Hist.* **17**, 391–399.

120. Waldron, H.A. (1978) Lead and human behaviour, *J. Ment. Defic. Res.* **22**, 69–78.

121. Waldron, H.A. and Stöfen, D. (1974) *Sub-Clinical Lead Poisoning.* Academic Press, London.

122. Zecevic, N. and Rakic, P. (1976) Differentiation of Purkinje cells and their relationship to other components of the developing cerebellar cortex in man, *J. Comp. Neurol.* **167**, 27–48.

123. Zook, B.C., London, W.T., Wilpizeski, C.R. and Sever, J.L. (1980) Experimental lead paint poisoning in non-human primates III Pathologic findings, *J. Med. Primatol.* **9**, 343–360.

Chapter 6

NEUROLOGICAL CONSEQUENCES OF MANGANESE IMBALANCE

Satya V. Chandra

1. INTRODUCTION

Manganese, a reddish-grey or silvery soft metallic element, was first recognized by Scheele and isolated by Gahn in 1774.[77] Manganese is said to be named after the Italian word for "magnesia." It is also thought to be named after the Latin word "magnes" (magnet) for the supposed magnetic properties of pyrolusite, the most common form of manganese mineral. Manganese is the twelfth most abundant element in the earth's crust and principal metallic component of nodules deposited on deep ocean floors. Recently, manganese nodules containing iron, nickel, copper, and cobalt have been discovered in Indian ocean.[90] Huge manganese ore deposits are found in the USSR, China, Brazil, India, Australia, Republic of South Africa, Ghana, and Morocco. Deposits of mineral ores containing less than 40% of manganese are also distributed in some parts of the United States of America. Manganese, in solution and in suspended material, was isolated from river water of USSR. The sediments of Rhine river and its estuaries in Europe have been found to contain manganese in concentrations varying from 1300

Satya V. Chandra: Industrial Toxicology Research Centre, Lucknow, India to 1800 ppm. Traces of manganese have also been detected in the waters collected from wells and springs in USA. High concentrations of manganese were reported in the well water from Japan and India.[57,59] A large number of plants contain manganese, the concentration of which is regulated by the metal content of the soil. In water and soil, manganese is found mostly in the diavalent form.

The most common form of manganese mineral is pyrolusite (MnO_2, black in color), which contains 60–63% of manganese. Other manganese minerals are psilomelane $BaMn^{2+}Mn_2^{+4}O_{16}(OH)_4$, black in color, hausmannite (Mn_3O_4), rhodocrosite ($MnCO_3$), rhodonite ($MnSiO_3$), and braunite ($3Mn_2O_3$)·$MnSiO_3$).[103] The percent manganese varies from 42 to 72 in these minerals. Manganese can exist in valences of 1, 2, 3, 4, 6, and 7; in organometallic complexes, manganese has valences of −3, −1, 0, and 5. The divalent and trivalent forms are the most stable, whereas the elemental form is a highly reactive metal. At least 300 minerals, including sulfides, anhydrous and hydrous oxides, carbonates, anhydrous and hydrous silicates, anhydrous and hydrous phosphates, arsenates, tungstates, borates, and so on contain manganese.[77]

Besides the natural occurrence of manganese, it is dispersed in the biosphere as a result of human activities. Ninety percent or more of the metal mined is used in the production of steel and cast irons. It is a constituent of manganese bronzes and other copper alloys. It provides strength, toughness, and hardness to steel. Manganese is also a component in Ni-Cr steels, which have wide application for their corosion and heat-resistant properties. It is utilized in the production of aluminum alloy, ferromanganese, and silicomanganese alloys. Manganese dioxide is used as an essential ingredient in the manufacture of common dry cell batteries. Manganese dioxide, as an oxidant, is used in the production of hydroquinone, in the leaching of uranium ores, and in the electrolytic production of zinc. In addition, manganese is used in the manufacture of various manganese compounds such as potassium permanganate, manganese sulfate, and manganese chloride. Manganese has been added as a trace element in fertilizers, particularly for soils deficient in manganese.

Manganese is also used in welding-rod coatings and fluxes. Various manganese salts are utilized to color or decolor glass and ceramic products. Manganese also finds use in the manufacture of pigments, paints, varnishes, fungicides, wood preservatives, and pharmaceuticals. Manganese oxides and powdered electrolytic metal are used in the production of manganese–zinc–ferrites in magnets for various electronic applications.

Recently a manganese tricarbonyl compound, commonly known as methyl-cyclopentadienyl manganese tricarbonyl (MMT), has been added to various fuels to improve combustion and as an octane improver in automobile gasolines. Thus release of manganese into the biosphere as a result of various human activities, has created great concern about possibly increasing health hazards from this metal.

2. METABOLISM OF MANGANESE

2.1. Absorption and Circulation

Manganese enters the metabolic system of animals through the gastro-intestinal and respiratory tracts. Entry through skin absorption is rather negligible. Common food items such as nuts, cereals, green leafy vegetables, fish, and other seafoods contain manganese in significant concentrations. However, the average dietary intake of manganese, usually in the form of organic chelates, is unlikely to contribute to an excess body burden of the metal in humans.

The average daily intake of manganese in human adults has been found to range from 3 to 9 mg, whereas the total body content of the metal is 10–20 mg.[29] Manganese intake in children fed breast milk is 11, 14, and 18 μg/d at 1, 3, and 5 months of age, respectively. However, children in the same age groups consuming cows milk ingest 24, 32, and 40 μg manganese/d, respectively.[2] The recommended average daily intake of manganese for children is 60 μg/kg body weight.[68] In animals, the ingestion of manganese varies enormously. Birds appear to have the highest requirement, and may reveal mild symptoms of manganese deficiency with diet containing 7–10 mg Mn/kg.[104]

Most of the data on the absorption of manganese is based on animal experimentation using inorganic salts of the metal. Absorption from the intestines of inorganic forms of manganese is slow and incomplete.[106] Administration of large amounts of manganese produced only 1% absorption and retention of the metal in the body tissues and 99% was excreted through the biliary passages in the intestines, eventually to be recovered in the feces. Oral administration of radiomanganese in experimental animals showed 97.2% recovery of radioactivity in the feces at the end of 75 h, indicating only a slight absorption through the intestines.[52]

Although the mechanism of manganese absorption is not well-defined, it appears to be absorbed throughout the length of the small intestines. Gastrointestinal tract plays a significant role even in the absorption of inhaled manganese. Mena et al.[70] demonstrated the appearance of radioactivity in the gastrointestinal tract in individuals exposed to radioactive manganese through inhalation. This suggested that inhaled manganese is removed from the respiratory tract by ciliary action and swallowed, eventually to be absorbed from the intestines. The percent absorption, however, depends on the particle size of the dust administered through the respiratory tract. Analysis of total body time/activity curves showed that total body retention in adult humans is only 0.21 ± 0.06% of ingested labelled manganese.[72] Injected labeled metal was completely retained in infant mice until 18 d after birth.[74] Cahill et al.[14] also observed 20 times more retention of manganese in infant rats compared to that found in adolescent or adult rats after oral administration of Mn_3O_4. In the young rat, intestinal absorption of manganese was 70% compared to 1–2% in the adults. Marked retention of manganese in the newborns was thought to result from incomplete maturation of the intestinal barriers to manganese.[71] Le Feure and Joel[64] believe that particulate manganese, in young animals is absorbed, dissolved, and transferred to portal blood through the pincytotic activity of intestinal epithelial cells; loss of this mechanism in adult animals permits the intestinal wall to act as a barrier against the uptake of particulate metal. Rehnberg et al.[88] suggested that several factors were responsible for decreased absorption of manganese from the intestines in adult animals: (1) cessation of pinocytosis; (2) change in the diet; (3) alterations in the activity of several enzymes in the intestinal mucosa; (4) morphological changes in intestinal mucosa; and (5) maturation of other physiological processes.

Several dietary constituents like calcium, phosphorus, iron, and soy proteins may interfere with the absorption and utilization of manganese[34,46,63,86,100] Individuals with iron deficiency anemia showed increased absorption of manganese as well as iron. The intestinal absorption of manganese in normal subjects was found to be 3 ± 0.5%, whereas in anemic patients, it was 7.5 ± 2.0%.[103] Alterations in iron absorption are paralleled by changes in the same direction in manganese absorption.[37] Experimental studies in iron-deficient rats have shown increased absorption and accumulation of manganese in body tissue compared to that observed in normal rats.[18] The plasma binding capacity of manganese by transferrin is increased by 100% and the entry of manganese into the brain is also increased significantly in anemic rats.[72]

It has also been observed that cobalt inhibits the uptake of manganese that has been transferred from intestinal mucosa to the circulation.[101]

Manganese, like iron, appears to be absorbed in a two-step mechanism; intestinal uptake followed by transfer across the mucosal surface into the circulation.[97] After being absorbed, manganese is bound to a specific plasma β-1-globulin transport protein, variously referred to as transmanganin[28] or transferrin.[60] It has been observed that transferrin molecules bind manganese and iron simultaneously. Recently it has been proposed that the transfer agent involved is metmanganoglobin, where Mn^{3+} can replace Fe^{3+}.[75] The metal binds to porphyrin molecules in the red blood cells.[11] The normal concentration of manganese in the whole blood of adults is 11 ng/mL while serum levels are 1–2 ng/mL. The concentration of manganese in the blood remains almost within normal limits after a manganese load indicating rapid clearance of the metal from the blood. According to reports in the literature, even heavy doses of manganese administered orally do not elevate blood levels of the metal and nearly 70% of the blood manganese leaves the circulation every minute.[103] Injected ^{54}Mn quickly disappears from the blood and concentrates in organs rich in mitochondria, e.g., the liver, pancreas, kidney, and pituitary. Manganese is highly concentrated in the mitochondrial fraction and also in the nucleus, where it is in a state of active exchange.[66]

Determination of the manganese content of adult human tissues reveals that a constant concentration of the metal is maintained throughout adult life, with the higest concentrations observed in the liver, pancreas, intestine, and kidneys (means of 68–180 ppm ash). Less than 25 ppm of manganese was noted in ash from aorta, heart, diaphragm, muscle, bladder, brain, esophagus, larynx, trachea, lung, spleen, skin, ovary, prostate, testes, uterus, thyroid, and bone. The bone content was only 2 ppm ash. Ash from adrenals, stomach, and omentum contained 36–48 ppm of manganese.[97] Hair accumulates manganese in high concentrations.[3] Higher concentrations of manganese are also observed in other pigmented portions of the body, like retina, pigmented conjuctiva and dark skin.[30,99] Recent studies have demonstrated a high rate of manganese accumulation in the tissues of the preweanling rat.[88]

2.2. Turnover and Excretion

Total body turnover of manganese is a function of the size of the manganese pool. The normal turnover in humans is of the order of 1 month; it is accelerated in miners exposed to manganese and in anemic patients.[103] Determination of the half-times of disappearance of injected [54]Mn in the whole body, liver, head, and thigh of normal individuals, healthy manganese miners, and patients with chronic manganese poisoning revealed an accelerated turnover in the actively working miners. The metal had a faster rate of turnover in the liver than in the head or thigh. The study indicated the existence of an expanded, rapidly exchanging manganese pool in the healthy miners compared to normal controls and patients with chronic manganese poisoning not recently exposed to manganese.[72] An elevated turnover of the metal is related to elevated tissue concentration, however, the latter does not correlate with the neurological manifestations of chronic manganese poisoning.[27]

The adult human body has an effective homeostatic mechanism for the maintenance of constant tissue concentrations of manganese that is based on controlled excretion rather than absorption. Endogenous manganese entering the intestines through biliary secretions is chiefly responsible for the regulation of manganese homeostasis. Manganese excreted through the biliary passage may again be absorbed through the intestines to become involved in the enterohepatic circulation, which also participates in the regulation of manganese excretion.[81] The concentration of manganese in the bile was found to be over 10 times greater than observed in the plasma.[12] This high bile/plasma concentration ratio and the high rate of biliary transport may suggest an active transport mechanism for the excretion of manganese into the bile.[62]

The work of Klaassen[62] refuted the earlier suggestion of Tichy and Cikrt[102] that the transfer of manganese from plasma into the bile occurs by a passive transfer mechanism, followed by a nonenzymatic complex formation in the bile. Whatever the mechanism may be, it is undoubted that the biliary route is very important for the excretion of manganese from the body. However, this route is not the only route for manganese excretion since manganese is also excreted in the feces even after biliary ligation. When the biliary route is blocked, manganese was found to be excreted into the intestine through pancreatic juice, indicating a role of the pancreas in the excretion of manganese.[103] Urinary excretion is rather negligible in normal persons; however, it may be increased by the administration of chelating agents. Injections of manganese sulfate ($MnSO_4$) in large doses followed by the administration of calcium ethylenediaminetetraacetate resulted in the presence of considerable amounts of the metal in the urine. These results suggested that stable manganese in the body may be available for chelation and diversion into the urine.[83] Normal urinary excretion of manganese ranges from 1.0 to 7 or 8 µg/L.[26] Ajemian and Whitman[1] showed that Mn urinary excretion in adults is 1 to 10 µg/L and that levels of manganese greater than 10 µg/L indicated industrial manganese exposure.

2.3. Metabolic Role

It has been established that a wide variety of organisms like bacteria, plants, and mammals require manganese for normal functioning of body tissues. Although the specific biochemical role of manganese in humans has not been defined, it has long been recognized as an activator of some enzymes that require the presence of a divalent ion. Such manganese-activated enzymes include hydrolases, kinases, decarboxylases, and transferases. Prolidase and succinic dehydrogenase are said to require manganese exclusively. Pyruvate carboxylase, a mitochondrial enzyme that catalyzes the carboxylation of pyruvate to form oxaloacetate is a manganese metalloenzyme.[77] Manganese has been shown to play a significant role in the utilization of glucose.[44,48] Manganese has also been associated with arginase, the activity of which has been found to diminish in the presence of manganese deficiency.[72] Manganese and magnesium are important metal ions associated with ATP, and manganese can replace magnesium in several biological reactions of the metal–ATP complex. Some early evidence suggested that manganese may be a cofactor for oxidative phosphorylation,[65] and it has been reported to be an activator in the synthesis of fatty acids and cholesterol in rats.[33] Manganese deficiency has been shown to produce defects in reproduction, although the precise biochemical role of this metal in reproduction has not been defined thus far.[77]

3. Manganese Deficiency and Neurological Disorders

Manganese has been classed among essential nutrients required for the normal functioning of living organisms. The daily human requirement of 1.2 mg of the metal is met by the presence of manganese in freely chosen diets and, as such, a deficiency of manganese has never been reported in human beings. Recently, manganese deficiency was noticed in a human subject receiving an experimental diet deficient in vitamin K. Manganese was not added in this diet and certain biochemical and hematological changes were observed that were corrected by manganese supplementation.[38] A deficiency of this metal can be produced experimentally in various species of animals by means of a manganese-deficient diet. The manifestations of manganese deficiency include retardation of growth, induction of bone abnormalities, dysfunction of the reproductive system, and disorders of the central nervous system. Birds maintained on a manganese-deficient diet develop bone deformities and dislocation of the Achilles tendon. The condition is commonly known as perosis. Offspring of these perotic birds showed symptoms of ataxia and the ataxic young of manganese deficient mothers have also been produced in other species of animals.[28] In rats, the offspring exhibited pronounced ataxia, lack of equilibrium, head retraction, and tremors.[56] Administration of a manganese-deficient diet to pregnant rats up to the 18th d of gestation resulted in the development of ataxia in almost all the surviving young. Four generations of rats raised on a diet supplying 0.03 mg of Mn/d/rat revealed symptoms of congenital ataxia, loss of equilibrium, backward gait, and postural abnormalities in subsequent generations of deficient animals.[54] The brains of these rats examined, under a light microscope with routine staining, did not reveal any abnormality. Shils and McCollum,[96] using a solid diet containing 20% powdered milk with 3 μg Mn/d/rat, observed ataxia, incoordination, and loss of balance in these animals. Manganese-deprived mice were unable to maintain orientation when submerged in water,[43] and ataxic rats and mice could not right themselves while swimming or falling in the air. A genetic mutant in mice, pallid, exhibited similar defects in coordination.[80] The young of manganese-deficient guinea pigs showed permanent damage, including head retraction, incoordination, and tremors throughout their life.[80] However, adequate supplementation of the metal in the diet during pregnancy prevented these disorders.

Miller et al.[73] noted the appearance of lameness and stiffness of legs in pigs maintained on a manganese-deficient diet. Detailed studies on manganese-deficient pigs later showed that their young ones were weak, ataxic, and displayed defects in locomotion and balance.[85] Poor growth, leg deformities, and poor fertility were also noted among cows grazing on manganese-poor pastures.[104]

Investigations of the mechanism of ataxia and incoordination in the offspring of manganese-deficient mothers revealed that the defects were due to the failure of otolith development in the maculae of the inner ear. Manganese deficiency impairs the development of the organic matrix containing acid mucopolysaccharides that is the main constituent of the otolithic membrane.[77]

A manganese-deficient diet given to the rats until the 20th d of pregnancy produced a fall in norepinephrine and dopamine concentrations in the brains of the mothers. The brains of their offspring also exhibited a deficiency in norepinephrine compared to the concentrations found in control animals. These observations show a direct link between the metabolism of catecholamines and manganese concentrations in the brain.[72]

4. Neurobehavioral Effects of Manganese Toxicity

4.1. In Humans

The neurological syndrome of chronic manganese poisoning was recognized, for the first time, by Couper[32] in five workers handling manganese dioxide (MnO_2) in a pyrolusite mill. The appearance of hypokinesia, akinesia, rigidity, tremor, and masklike facies in the patients with chronic manganese intoxication were confirmed subsequently by Emden[42] and Jaksch.[58] Since then, cases of chronic manganese encephalopathy continue to be the subject of reports from various parts of the world including the USSR, Morocco, Chile, India, Japan, and the USA. The highest incidence of manganese poisoning was reported among workers exposed to this metal in Chile.[72] In India, the prevalence has been estimated to be as high as 25%.[87] The sources of manganese poisoning include a wide variety of operations, such as mining, ore crushing, manufacture of ferromanganese alloys, steel, dry cell batteries, various manganese-based chemicals, and through arc welding. Although the toxicity most commonly occurs by inhalation of the metal dust or fumes, cases of manganese poisoning have been reported in the general population ingesting manganese through drinking water from a well near which manganese-containing batteries had been buried.[59] A case of extrapyramidal neurological disease was reported in a person who had consumed a large amount of multimineral tablets. An increased manganese concentration was found in the brain, serum and hair of this patient.[5]

The characteristic signs and symptoms of manganism have been described in detail by Flinn et al.,[45] Rodier,[89] Penalver,[83] Tanaka and Lieben,[98] Chandra et al.,[19] and Cook et al.[27] According to the most elegant description of Rodier,[89] the signs and symptoms of poisoning could be divided into three phases: the prodromal period, the intermediate phase, and the established phase. The onset is usually insidious, beginning with the prodromal phase, which is usually termed as "manganese psychosis." The symptoms of apathy, asthenia, and anorexia follow unaccountable laughter, euphoria, impulsiveness, insomnia, and muscular pains. Sometimes, sexual excitement follows impotence. The symptoms of "manganese madness" were invariably the first to appear in cases reported from Chile, characterized by hallucinations, delusions, and compulsions. These symptoms usually lasted for 1–3 months.[72] Following or concomitant with these symptoms, the intermediate phase begins with disturbances of speech characterized by slow and difficult articulation, incoherence, expressionless facies, general clumsiness of movements, altered balance, and altered gait. The reflexes in lower limbs become exaggerated and the patient is unable to climb or descend a ladder. The established neurological phase presents with the signs of muscular rigidity and slow, spasmodic, and staggering gait, commonly referred as "cock walk." Fine tremors usually affecting the upper limbs are noticed, and these are exaggerated by fatigue or emotion. Spasmodic laughter and excessive sweating are also noticed in some patients. A summary of the significant neurological disorders in various grades of manganese poisoning is shown in Table 1. Despite the severity of the neurological symptoms, the patient survives, although permanently disabled.

Table 1
Summary of Significant Neurological Disorders in Various Grades of Manganese Poisoning

Grades of poisoning	Significant neurological disorders
I. Mild (manganese psychosis)	Asthenia, anorexia, insomnia, muscular pains, mental excitement, hallucinations, unaccountable laughter, impaired memory, compulsive actions, and muscular pains
II. Moderate	Speech disorders, clumsiness of movements, abnormal gait, altered balance, exaggerated reflexes in lower limbs, expressionless facies, adiodokinesis, and fine tremors
III. Severe	Tremors, "cock walk," rigidity in both the extremities, spasmodic laughter

Cases presenting with similar signs and symptoms of neurobehavioral disorders were reported from the manganese-based industries, such as the German fertilizer industry using $MnSO_4$ in fertilizer and Czech plants manufacturing permanganate. Cases of manganese poisoning among miners and arc welders were also reported from other countries including India.[4,25,79,108]

In most of these studies, the presymptomatic exposure period averaged 6 months to 2 yr with a range of 1 month to 10 yr. This great variability of time exposure to develop the disease may be caused by the frequent changes in jobs, variation in natural ventilation, and fluctuations in the concentration of the metal dust in the work place. Certain metabolic and social factors have been implicated as likely to render an individual susceptible to manganese encephalopathy. Lesions of the excretory systems (liver, kidney) may be of significance in view of the slow elimination of the metal from the body. The ability of the respiratory system to eliminate manganese may also determine the individual susceptibility. Alcoholism has been implicated in determining the individual susceptibility to manganese poisoning; chronic infections such as syphilis, malaria, tuberculosis, and avitaminosis are also believed to influence the sensitivity of the individual in the development of manganese toxicity. Another possible factor affecting susceptibility is an increased intestinal absorption of manganese, e.g., individuals with iron deficiency or cirrhosis have an increased rate of intestinal absorption of both manganese and iron.[72] Although the role of manganese as a weak chemical allergen has been mentioned in inducing respiratory illness, it is difficult to assess the significance of this factor in rendering an individual susceptible to neurological disease resulting from manganese exposure.[97]

4.2. In Animals

Most experimental studies concerning the neurotoxic effects of manganese have utilized small laboratory animals: however, these are of limited use in evaluating the neurobehavioral manifestations of extrapyramidal neurological disorders. Mella in 1924[69] exposed rhesus monkeys to manganese chloride ($MnCl_2$) and demonstrated the development of choreic or choreo-athetoid type of movements after 18 months of exposure. These monkeys also exhibited rigidity and fine tremors in the hands. Neuropathological examination revealed widespread neuronal degeneration and neuroglial proliferation. An elegant model of manganese neurotoxicity was produced in monkeys exposed through inhalation to a metal ore dust containing 50% MnO_2[105] The animals developed ataxia, widebased gait, intention tremor, and paralysis of the hind limbs. Histological examination showed degeneration of Purkinje cells and a partial disintegration of the granules over wide areas of the cerebellum. Pentschew et al.[84] administered MnO_2 intramuscularly and elicited the symptoms of excitability and clumsy movements in monkeys. The consequences of subcutaneous administration of 200 mg of MnO_2 to squirrel monkeys at monthly intervals were concomitant symptoms of rigidity, and fine tremors; these symptoms were noticed 2 months after the first injection. Despite marked neurochemical changes, no histological damage was observed in any of the brain regions.[78] Furthermore, not all exposed monkeys developed the signs and symptoms of neurological involvement at the same time after the injection, which indicated that other factors also had a role in determining the susceptibility of an animal to manganese intoxication. In another study, intratracheal inoculation of rabbits with MnO_2 produced biochemical as well as histopathological changes in the brain.[76] The histological changes were in the form of neuronal loss in the cerebral and cerebellar cortices, caudate nucleus, putamen, and substantia nigra.[17]

Most of these studies deal with neuropathological changes during chronic manganese intoxication: however, recently, scattered neuronal degeneration in the regions of the cerebral and cerebellar cortices have been demonstrated in rats and rabbits after a short-term exposure to the metal.[15,16] Furthermore, these alterations could be observed in the brain of growing animals much earlier than in adults.[20] In smaller animals, the symptoms of early manganese poisoning can be detected only with special procedures, particularly behavioral studies. However, in the chimpanzee, early manganese toxicity is manifested in the form of difficulty in climbing or descending a fence, athetoid extension, dystonic posturing, mask-like face, and somnolence.[77]

Manganese encephalopathy could be produced in various species of animals by the administration of inorganic forms of the metal and after exposure to an organic compound, methylcylopentadienyl tricarbonyl manganese (MMT). The threshold of neuromuscular excitability was lowered in animals exposed to this compound, indicating increased sensitivity. Inhalation of manganese tricarbonyl compounds produced an increase in the threshold of neuromuscular excitability 7 months after the exposure of rats.[77] A series of studies of repeated inhalation exposures of this compound by mice, rats, guinea pigs, rabbits, cats, and dogs showed increased mortality in rats and mice, but not in the other species. Signs of toxicity included mild excitement, hyperactivity, tremors, severe spasms, slow and labored respiration, mild clonic convulsions, and terminal coma. These experimental results led the investigators to propose a threshold limit value of 0.2 mg/m^3 or 0.1 ppm expressed as manganese for exposure.[77] However, the contributions of organometallic manganese fuel additives to the problem of air pollution and related health hazards in exposed populations requires considerable further study in order to recommend threshold limit values for such compounds.

5. Neurotransmitters in Manganese Intoxication

There is sufficient experimental evidence to indicate that neurochemical changes precede the neuropathological alterations in manganese toxicity. The symptoms and signs of manganese encephalopathy share several features in common with Parkinson's disease in which decrease of melanin in the regions of the substantia nigra and locus ceruleus is accompanied by the depletion of dopamine (DA) and probably precedes structural changes in the brain.[72] Depletion in the contents of brain DA were also reported in a patient who had suffered from chronic manganese poisoning.[55] Thus the neurochemical studies performed in cases of Parkinson's disease have been duplicated with experimental manganese poisoning to unravel the pathogenesis of manganese neurotoxicity.

Papavasiliou and colleagues[82] demonstrated an intracellular increase in the concentration of manganese in liver tissue after the administrations of DA, L-epinephrine, or DL-isoproterenol. It was also observed that this effect was mediated through an intracellular increase in cyclic 3',5'-adenosine monophosphate. Although the importance of these studies in relation to manganese encephalopathy could not be evaluated, it became clear that a correlation existed between manganese metabolism and biogenic amines.

A direct relationship between brain biogenic amines and manganese was demonstrated for the first time by Neff and colleagues[78] using squirrel monkeys repeatedly exposed to MnO$_2$. These monkeys exhibited postural abnormalities, rigidity, and tremors. The levels of DA and serotonin (5-HT) were markedly reduced in the caudate nucleus without any morphological changes in the brain. The magnitude of DA depletion correlated well with the degree of neurotoxicity as assessed by extrapyramidal neurological disorders. In another study, rabbits were inoculated, intratracheally, with MnO_2 and the development of neurological disorders was observed up to 24 months. At the termination of the experiment, whole brain DA and norepinephrine (NE) contents were reduced by 21 and 58%, respectively, without any change in 5-HT.[76] Cerebral levels of 5-HT were depressed in the brains of rats fed a diet containing 2 mg manganese.[61] Bonilla and Diez-Ewald[8] observed decreased levels of DA and homovanillic acid (HVA) in the brain of rats chronically exposed to manganese. The administration of L-dopa to control and manganese-exposed rats produced an increase in both DA and HVA concentrations to the same levels, indicating that the metal does not directly affect dopa decarboxylase, monoamine oxidase (MAO), or catechol-o-methyltransferase (COMT) to produce reduction in the brain DA after chronic manganese intoxication. These observations led the authors to propose that the depletion of DA in chronic manganese intoxication may be caused by inhibition of tyrosine hyroxylase (TH). In an attempt to understand the mechanism of striatal DA deficiency, Donaldson et al.[39] investigated Na^+K^+-ATPase in microsomal preparations from rat brain regions, considering that the depletion of DA could result from the impaired transport of L-dopa across the neuronal membrane because of inhibition of transport ATPase. However, manganese had no such effect.

Deskin et al.[35] reported the highest concentration of manganese and neurochemical changes only in the region of the hypothalamus in neonatal rats chronically exposed to manganese. The maximum accumulation of manganese in this region compared to that observed in the corpus striatum of manganese-exposed animals and may result from a rich blood supply to the hypothalamic area.[40] Deskin et al.[35] observed a significant reduction in the endogenous contents of DA in the hypothalamic area of rats chronically exposed to manganese; however, the depletion of DA after administration of α-methyl-p-tyrosine was less in this region in manganese-treated rats, indicating a reduction in DA turnover. A significant decrease in the activity of TH in the hypothalamic area of manganese-exposed rats is likely to be responsible for the decrease in the synthesis of DA. In the same experiment, NE levels in the hypothalamic area remained unaltered after chronic manganese administration to rats, thus ruling out the possibility of NE feedback inhibition as a mechanism to reduce the activity of TH.[107] Tyrosine hydroxylase activity has been found to be increased in neostriatum, midbrain, and hypothalamus of rats orally treated with manganese chloride for 3 months, but after 8 months of manganese administration, a significant decrease was observed in the enzymatic activity in neostriatum only.[10] However, from these studies it was difficult to determine whether manganese directly affects TH activity or whether it has an effect at the level of the synapse. In vitro studies by Deskin and coworkers[36] showed that manganese does not directly inhibit TH activity; however, the possibility that manganese interferes with the ability of TH to be mebrane-bound is not ruled out as a mechanism for reducing the activity of TH. The same group of investigators also ruled out the possibility that competition between manganese and other physiologically important divalent cations might cause the reduction in activity of TH seen in vivo after chronic manganese administration. Whether the trace element shift in the brain observed during manganese intoxication is responsible for inhibition of the activity of TH is unresolved at present.[21] Manganese, as the divalent ion, is known to stimulate catecholamine oxidation.[49,50] Donaldson et al.[41] reported autoxidation of DA in the presence of Mn^{2+} with a concomitant increase in free-radical production. Manganese, by increasing DA autoxidation, could augment the cytotoxins emanating from this process and thus could contribute to the neurodegenerative changes. Increase in striatal γ-aminobutyric acid (GABA) has also been reported in rats chronically exposed to manganese,[9] and this may indicate a role for GABAergic function in the regulation of the dopaminergic system in chronic manganese intoxication. The effects of manganese exposure on biogenic amines in the brain are shown in Table 2.

In an attempt to see whether manganese exerts its neurotoxic effect mainly by affecting the dopaminergic system, an assay of receptor–ligand interaction was conducted to detect derangement of a specific neurotransmitter system.[92] A significant increase of striatal spiroperidol binding occurred in manganese-exposed animals. At higher dose levels of manganese, cerebellar GABA, frontal cortical 5-HT, and striatal muscarinic binding were depressed. However, the striatal levels of enkephalin, substance P, DA, 5-HT, or DOPAC remained unaltered in this experiment.

Changes in the concentration of various neurotransmitters in the brain of chronically exposed animals led the investigators to study the effect of manganese on various enzymes involved in the metabolism of biogenic amines. Rats treated with $MnCl_2·4H_2O$ (8 mg/kg/d) intraperitoneally for a period of 120 d showed a significant reduction in the activity of acetylcholine esterase in cere-

Table 2

Effects of Manganese Exposure on Biogenic Amines in the Central Nervous System of Animals

References	Species	Manganese administration	Region of brain	Effects observed
Neff et al.[78]	Squirrel monkeys	200 mg MnO_2, subcutaneously at monthly intervals	Caudate nucleus	Dopamine and serotonin depletion, 2 months after exposure
Mustafa and Chandra[76]	Rabbits	400 mg MnO_2, intratracheally	Whole brain	Dopamine and norepinephrine depletion, 24 months post innoculation
Bonilla and Diez-Ewald[8]	Rats	5 mg $MnCl_2$/mL drinking water, daily	Cerebral hemispheres including basal ganglia	Dopamine depletion, 7 months after exposure
Deskin et al.[35]	Rats	20 μg/g $MnCl_2 \cdot 4H_2O$ orally, daily	Hypothalamus	Dopamine depletion, 25 days after exposure
Bonilla[9]	Rats	10 mg $MnCl_2$/mL in drinking water, daily	Caudate nucleus	Gamma-aminobutyric acid elevation, 2 months after exposure
Cotzias et al.[31]	Mice	1.0 mg ^{55}Mn/mL in milk diet	Cerebrum	Dopamine elevation, 80 d after exposure
Chandra et al.[22]	Mice	Pups exposed from birth through milk of mothers receiving 5 mg/mL $MnCl_2$ in drinking water. After weaning at 25 d, exposed directly to 3 μg Mn^{2+}/mL in drinking water	Corpus stratium	Dopamine and norepinephrine elevation, 30 d after exposure

brum, cerebellum, and the rest of the brain. The activity of monoamine oxidase (MAO) was increased in the cerebellum only.[95] Several reports of increased MAO activity in the brain during manganese intoxication[20,91,93] indicate the physiological importance of this enzyme in relation to the levels of brain biogenic amines during manganese intoxication. Both manganese and MAO are concentrated in the mitochondria,[28] thus manganese may have a direct effect on the activity of this enzyme and consequently on the changes in the levels of biogenic amines. However, increased MAO activity does not appear always to coincide with the decreased DA concentrations.

Manganese is normally concentrated in melanin[31] and additional accumulation of the metal during chronic manganese intoxication could be deleterious. Histochemical studies in the brains of monkeys chronically exposed to manganese revealed loss of melanin granules in the region of substantia nigra.[53] Whether this loss of melanin was caused by the degeneration of neurons or of manganese-induced disturbances in the metabolism of catecholamines resulting in the reduced synthesis of melanin, could not be ascertained. It is also not clear whether depletion of melanin precedes or is secondary to the impairment in the metabolism of brain catecholamines.

In summary, chronic manganese administration to experimental animals resulted in neurological disorders of extrapyramidal involvement and neurochemical changes in the region of the corpus striatum and hypothalamus. The genesis of neurochemical changes in chronic manganese intoxication will be discussed after describing the effect of short-term exposure to manganese on the profile of neurotransmitters in the central nervous system. Although Neff et al.[78] are credited with being the first to demonstrate the significance of brain catecholamines in chronic manganese intoxication, it was the work of Chandra and colleagues.[22-24] that provided a possible neurochemical mechanism for early manganese encephalopathy whereby the disease process might be reversed through suitable preventive and therapeutic measures. Radical differences in manganese metabolism in postnatal or neonatal versus adult animals[31] led the investigators to use growing animals for their experiments, hopefully to provide a basis for identifying and studying effects of manganese excess early in human life. Suckling rats were exposed to manganese through the milk of nursing dams receiving 15 mg $MnCl_2 \cdot 4H_2O$/kg/d. No significant difference was observed in the growth rate, developmental landmarks, or walking pattern in manganese-exposed

pups compared to controls during the entire 30-d period of exposure. However, metal accumulation and significant enzymic alterations in the brain of exposed pups were more pronounced compared to their nursing dams. These results indicated that the
growing brain was more susceptible to the neurotoxic effects of
manganese than were those of adults.[91] Neonatal mice and rats
do not excrete manganese for the first 17–18 d of life, although absorption and tissue accumulation is extensive. Further, rats
younger than 18 d of age are susceptible to a greater penetration of
manganese into the brain.[74]

Behavioral and neurochemical studies were conducted in
mice exposed to manganese from birth through the milk of
manganese-exposed mothers and after weaning directly through
drinking water containing 3 µg Mn^{2+}/mL.[22] Motor activity of
manganese-exposed offspring increased significantly at 60 and 90
d and was associated with significant elevation in the levels of tyrosine, DA, NE, and HVA in the corpus striatum.[22] Increased levels
of brain DA and NE are known to produce locomotor activation.[7,13] Furthermore, Goodwin and Sack[51] have suggested that
an excess of NE accompanies mania. Therefore, increased locomotor activation caused by the elevation in the striatal levels of
catecholamines during early manganese intoxication can be correlated with the manifestations of psychiatric disturbances observed
during the prodromal period of manganese poisoning in humans.
To understand the mechanism of manganese-induced elevation of
brain catecholamines, their synthesis and rate of decline were determined by using 3,5-³H-tyrosine and α-methyl-p-tyrosine, respectively, in growing animals.[23] The synthesis of labeled
catecholamines in the brain at 60 min after the administration of
3,5-³H-tyrosine was more pronounced in manganese-exposed rats
than in controls (Fig. 1). The same manganese treatment produced
significant elevation in the levels of HVA in the region of corpus
striatum. Assuming that an increase in HVA is indirect evidence of
enhanced DA synthesis,[6] the observation of increased striatal content of HVA after manganese treatment supports the view that a
short-term exposure to manganese causes enhanced synthesis of
brain catecholamines.[22] Furthermore, the turnover rates of
catecholamines after administration of α-methyl-p-tyrosine, a potent inhibitor of tyrosine hydroxylase, were also found to be significantly higher in manganese-exposed rats compared to their
counterpart controls.[23] Iron is known to stimulate the activity of
tyrosine hydroxylase,[67] and since the metabolism of iron and
manganese has certain similarities,[100] it may be reasonable to as-

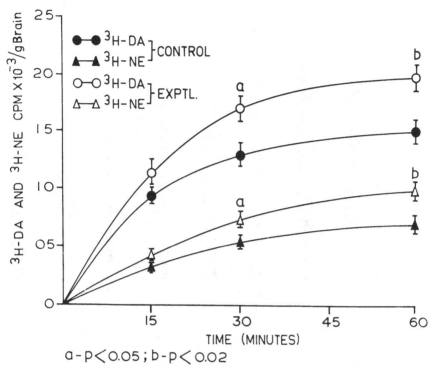

FIG. 1. Effect of manganese on the synthesis of ³H-dopamine and
³H-norepinephrine in the brain.

sume that manganese may also activate tyrosine hydroxylase to
produce enhanced biosynthesis of brain catecholamines during
early manganese intoxication. The brain contents of NE increased
to a greater magnitude than DA, possibly indicating a manganese-
induced increase in the activity of dopamine β-hydroxylase. Experiments have demonstrated significant increase in the contents
of brain copper during early manganese intoxication.[94] Since copper is known to activate dopamine β-hydroxylase,[47] any increase
in the activity of this enzyme may be caused by increased copper
concentration rather than the direct effect of manganese. However, investigations on the effect of manganese on the activity of
enzymes involved in the biosynthetic pathway of biogenic amines
are needed to understand the mechanism of increased turnover of
brain catecholamines during early manganese intoxication.

Experiments on the effect of manganese on the levels of brain
catecholamines revealed major differences during short and long-
term exposure to the metal. Chandra and Shukla[24] investigated
the levels of various neurochemicals in rats exposed to manganese

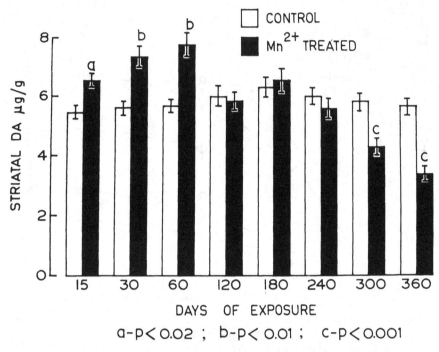

a-p< 0.02 ; b-p< 0.01 ; c-p< 0.001

FIG. 2.   Effect of manganese on the striatal content of dopamine.

through drinking a solution containing $MnCl_2.4H_2O$ (1 mg/mL) in
water, at different time intervals up to a period of 360 d. Manganese was found to induce an elevation in the levels of striatal DA
and NE up to a period of 60 and 120 d, respectively. The levels declined at 180 and 240 d and a significant depletion in the striatal
contents of DA and NE was noticed at 300 and 360 d after manganese exposure (Fig. 2). The variation and final reversal of effects on
the contents of striatal biogenic amines over time during chronic
manganese administration were not related to the accumulation
of this metal in the corpus striatum. Almost stationary levels of
manganese after a period of 240 d may result from the saturation
of binding sites for the metal (Fig. 3). It was concluded that manganese administration to the rats resulted in an initial elevation in
the contents of striatal catecholamines, which may be responsible
for early behavioral changes in the form of locomotor activation. A
similar neurochemical mechanism may also be responsible for
producing psychiatric illness during early manganese intoxication
in humans. The continued presence of excess manganese in the
brain affects other biomolecules,[93] particularly those involved in
oxidative phosphorylation, and results in disruption of metabolic

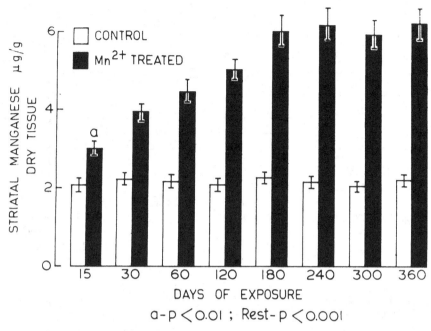

Fig. 3. Effect of manganese intake on the manganese content in the corpus striatum.

functions, breakdown of cellular structure, and degeneration of neurons. These changes, particularly in the region of the basal ganglia, may lead to decreased synthesis of catecholamines. Thus impairment in the dopaminergic response may produce symptoms of akinesia. As the disease progresses, severe neuronal loss and marked depletion in the chemical mediators may be responsible for development of the neurological disorders observed during chronic manganese intoxication. A possible mechanism of manganese encephalopathy has been illustrated in Fig. 4. In summary, animal studies indicate that a major neurochemical abnormality in manganese toxicity may arise from a defect in the metabolism of chemical mediators, which may in turn be responsible for the production of manganese encephalopathy.

6. Conclusions

Manganese, one of the most abundant elements in the earth's crust, is present in nearly all living organisms. The principal uses of manganese are (1) manufacture of steel, ferrous, and non-

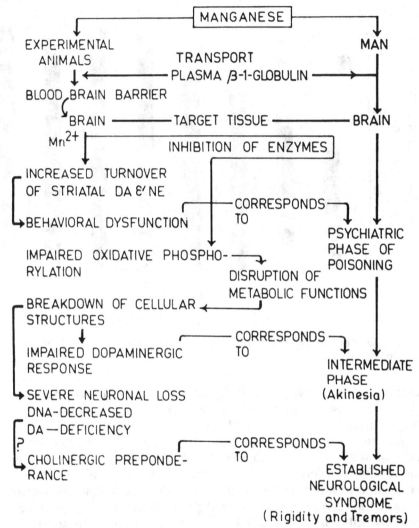

FIG. 4. Possible pathogenetic mechanism of manganese encephalopathy.

ferrous alloys, (2) manufacture of dry cell batteries, (3) in various chemical industries, and (4) in fuel oils.

Manganese is an essential trace element and the human daily requirement is 1.2 mg, which is met by the consumption of common food items such as nuts, cereals, green leafy vegetables, and fish. The main absorption route is the gastrointestinal tract, and even inhaled manganese is transferred to the intestine eventually to be absorbed. In the trivalent state, manganese is bound to a specific plasma β-1-globulin, commonly referred to as transmanganin or transferrin. The metal also incorporates into the porphyrin molecule of red blood cells. Normal manganese concentration in the whole blood of adult humans is around 11 ng/mL. About 70% of the normal blood manganese leaves the circulation every minute either to be deposited in tissues rich in mitochondria such as pancreas, kidneys, and liver or to be excreted through the feces. Manganese homeostasis in the body is regulated primarily by excretion rather than absorption. It is excreted almost exclusively into the intestinal tract via the biliary passage. Urinary excretion is rather negligible.

A deficiency of manganese can be produced experimentally in various species of animals by providing a manganese-deficient diet. The neurological consequences of manganese deficiency include ataxia, impaired equilibrium, head retraction, and tremors. Manganese-deprived mice are unable to maintain orientation when submerged in water, and ataxic rats and mice are unable to right themselves while swimming or falling in the air. The brains of offspring from manganese-deficient mothers exhibit a deficiency of norepinephrine compared to the amine level in control animals.

Chronic manganese neurotoxicity was first described by Couper in 1837 and has since been recognized as an occupational hazard in manganese-based industries. The symptoms and signs of manganese neurotoxicity appear after variable periods of exposure ranging from 6 months to 24 yr. The psychotic symptoms, characterized by hallucinations, delusions, unaccountable laughter, and euphoria, usually precede the neurological syndrome, and last for 1–3 months. Towards the end of manganese psychosis, neurological symptoms of involvement of the extrapyramidal system emerge. These symptoms include speech difficulty, bradykinesia, rigidity, difficulty in walking, impaired postural stability, tremors, and masked facies. Factors that may render an individual susceptible to manganese poisoning include iron deficiency, lesions of the excretory system, alcoholism, and chronic infections.

Experimental studies designed to elucidate the mechanism of manganese neurotoxicity reveal neuronal degeneration in the regions of basal ganglia, cerebral cortex, and cerebellar cortex. Loss of melanin granules is noticed in the region of substantia nigra in monkeys chronically exposed to manganese. The dopmaine, norepinephrine, and serotonin contents are reduced in the brain of experimental animals chronically intoxicated with manganese. Studies after short-term exposure to manganese reveal significant elevation in the levels of catecholamines in the region of the corpus striatum of rats and mice. A marked increase in the striatal levels of catecholamines after short-term manganese exposure is caused by increased synthesis of these amines, as measured by the amount of conversion of 3,5-^{3}H-tyrosine to ^{3}H-labeled catecholamines and the rate of decline of endogenous DA following inhibition of synthesis after administration of α-methyl-*p*-tyrosine. These neurochemical changes may be responsible for the production of behavioral aberrations in manganese-exposed animals. Thus it is hypothesized that the early psychiatric phase of manganese poisoning in humans may be the result of increased turnover of striatal catecholamines.

In comparison to adults, neonatal and growing brain is more susceptible to the toxic effects of manganese. Neonatal animals do not excrete manganese for the first 17–18 d of life, although absorption and tissue accumulation is extensive. Furthermore the factor of immaturity of the intestinal barrier to manganese absorption in premature and newborn infants may increase the risk of toxicity to this population group arising from increasing environmental pollution from manganese.

References

1. Ajemian, R. S. and Whitman, N. E. (1969) Determination of manganese in urine by atomic absorption spectrometry, *Amer. Ind. Hyg. Assoc. J.* **30,** 52–56.
2. Alexander, F. W., Clayton, B. E. and Delves, H. T. (1974) Mineral and trace metal balance in children receiving normal and synthetic diets, *Quart. J. Med.* **169,** 89–105.
3. Arkhipova, O. G., Tolgskaya, M. S. and Kochetkova, T. A. (1966) Toxic properties of manganese cyclopentadienyltricarbonyl antiknock substance, USSR literature on air pollution and related disease **12,** 85–89.
4. Balani, S. G., Umarji, G. M., Bellare, R. A. and Merchant, H. C. (1967) Chronic manganese poisoning, *J. Postgraduate Med.* **13,** 116–122.
5. Banta, G. and Markeshbery, W. R. (1977) Elevated manganese levels associated with dementia and extrapyramidal signs, *Neurology* **27,** 213–216.
6. Bariletto, S., Dollar, E. and Leitz, F. (1975) Effect of amantadine on the rate of dopamine synthesis in rat corpus striatum, *J. Neurochem.* **25,** 139–142.
7. Bartholini, G., Blum, J. E. and Pletscher, A. (1969) DOPA induced locomotor stimulation after inhibition of extracerebral decarboxylase, *J. Pharm. Pharmacol.* **21,** 297–301.
8. Bonilla, E. and Diez-Ewald, M. (1974) Effects of L-dopa on brain concentrations of dopamine and homovanillic acid in rats after chronic manganese chloride administration, *J. Neurochem.* **22,** 297–299.

9. Bonilla, E. (1978) Increased GABA content in caudate nucleus of rats after chronic manganese administration, *J. Neurochem.* **31**, 551–552.

10. Bonilla, E. (1980) L-tyrosine hydroxylase activity in the rat brain after chronic oral administration of manganese chloride, *Neurobehav. Toxicol.* **2**, 37–41.

11. Borg, D. C. and Cotzias, G. C. (1958) Incorporation of manganese into erythrocytes as evidence for a manganese porphyrin in man, *Nature* **182**, 1677–1678.

12. Brauer, R. W. (1959) Mechanism of bile secretion, *J. Amer. Med. Assoc.* **169**, 1462–1466.

13. Broitman, S. T. and Donoso, A. O. (1971) Locomotor activity and regional brain noradrenaline levels in rats treated with prenylamine, *Experientia* **27**, 1308–1309.

14. Cahill, D. F., Bercegeay, M. S., Haggerty, R. C., Gerding, J. E. and Gray, L. E. (1980) Age related retention and distribution of ingested Mn_3O_4 in the rat, *Toxicol. Appl. Pharmacol.* **53**, 81–91.

15. Chandra, S. V. and Srivastava, S. P. (1970) Experimental production of early brain lesions in rats by parenteral administration of manganese chloride, *Acta Pharmacol. Toxicol.* **28**, 177–183.

16. Chandra, S. V. and Sur, R. N. (1971) Early brain changes in rabbits induced manganese chloride. *Environ. Res.* **3**, 417–424.

17. Chandra, S. V. (1972) Histological and histochemical changes in experimental manganese encephalopathy in rabbits, *Arch. Toxikol.* **29**, 29–38.

18. Chandra, S. V. and Tandon, S. K. (1973) Inhanced manganese toxicity in iron deficient animals, *Environ. Physiol. Biochem.* **3**, 230–235.

19. Chandra, S. V., Seth, P. K. and Mankeshwar, J. K. (1974) Manganese Poisoning: Clinical and biochemical observations, *Environ. Res.,* **7**, 374–380.

20. Chandra, S. V. and Shukla, G. S. (1978) Manganese encephalopathy in growing rats, *Environ. Res.* **15**, 28–37.

21. Chandra, S. V., Srivastava, R. S. and Shukla, G. S. (1979) Regional distribution of metals and biogenic amines in the brain of monkeys exposed to manganese, *Toxicol. Lett.* **4**, 189–192.

22. Chandra, S. V., Shukla, G. S. and Saxena, D. K. (1979) Manganese induced behavioral dysfunction and its neurochemical mechanism in growing mice, *J. Neurochem.* **33**, 1217–1221.

23. Chandra, S. V. and Shukla, G. S. (1981) Effect of manganese on synthesis of brain catecholamines in growing rats, *Acta Pharmacol. Toxicol.* **48**, 349–354.

24. Chandra, S. V. and Shukla, G. S. (1981) Concentrations of straital catecholamines in rats given manganese chloride through drinking water, *J. Neurochem.* **36**, 683–687.

25. Chandra, S. V., Shukla, G. S., Srivastava, R. S., Singh, H. and Gupta, V. P. (1981) An exploratory study of manganese exposure to welders. *Clin. Toxicol.* **18**, 407–416.

26. Cholak, J. and Hubbard, D. M. (1960) Determination of manganese in air and biological material. *Amer. Ind. Hyg. Assoc. J.* **21**, 356–360.

27. Cook, D. G., Fahn, S. and Brait, K. A. (1974) Chronic manganese intoxication. *Arch. Neurol.* 30, 59–64.
28. Cotzias, G. C. (1958) Manganese in health and disease. *Physiol. Rev.* 38, 503–532.
29. Cotzias, G. C. (1962) In *Mineral Metabolism,* C. L. Comar and F. Bronner (Eds.) Vol. 2, Academic Press, New York, pp. 403–442.
30. Cotzias, G. C., Papavasiliov, P. S. and Miller, T. S. (1964) Manganese in melanin. *Nature* 201, 1228–1229.
31. Cotzias, G. C., Papavasiliou, P. S., Mena, I., Tang, L. C. and Miller, S. T. (1974) Manganese and catecholamines. In: F. H. McDowell and A. Barbeau (Eds.), *Advances in Neurology,* Vol. 5, Raven Press, New York, pp. 235–243.
32. Couper, J. (1837) On the effects of black oxide of manganese when inhaled into the lungs. *Br. Ann. Med. Pharmacol.* 1, 41–42.
33. Curran, G. L. (1954) Effect of certain transition group elements on hepatic synthesis of cholesterol in the rat, *J. Biol. Chem.* 210, 765–770.
34. Davis, P. N., Norris, L. C. and Krantzer, F. H. (1962) Interference of soybean meal with the utilization of trace minerals. *J. Nutr.* 77, 217–223.
35. Deskin, R., Bursian, S. J. and Edens, F. W. (1980) Neurochemical alterations induced by manganese chloride in neonatal rats. *Neurotoxicology,* 2, 65–73.
36. Deskin, R., Bursian, S. J. and Edens, F. W. (1980) An investigation into the effects of manganese and other divalent cations on tyrosine hydroxylase activity. *Neurotoxicology* 2, 75–81.
37. Diez-Ewald, M., Weintraub, L. R. and Crosby, W. H. (1968) Interrelationship of iron and manganese metabolism. *Proc. Soc. Exp. Biol. Med.* 129, 448–451.
38. Doisy, E. A., Jr. (1974) Effects of deficiency in manganese upon plasma levels of clotting proteins and cholesterol in man. In: W. G. Hoekstra (Ed.), *Trace Element Metabolism in Animals,* Proceedings of the 2nd international symposium, University Park Press, Baltimore, pp. 668–670.
39. Donaldson, J., Cloutier, T., Minnich, J. L. and Barbeau, A. (1974) Seizures in rats associated with divalent cation inhibition of Na^+,K^+-ATPase, *Canad. J. Biochem,* 49, 1217–1222.
40. Donaldson, J., Cloutier, T., Minnich, J. L. and Barbeau, A. (1974) Trace metals and biogenic amines in rat brain. *Adv. Neurol.* 5, 245–252.
41. Donaldson, J., LaBella, F. S. and Gresser, D. (1980) Enhanced autooxidation of dopamine as a possible basis of manganese neurotoxicity. *Neurotoxicology* 2, 53–64.
42. Emden, H. (1901) Veber eine Nervenkrankheit nach Manganvergiftung. *Munch. Med. Wochenschr.* 48, 1852–1853. Quoted from NAS, Manganese, 1973.

43. Erway, L., Hurley, L. S. and Fraser, A. (1966) Neurological defect: Manganese in phenocopy and prevention of genetic abnormality of inner ear. *Science* **152**, 1766–1768.

44. Everson, G. J. and Shrader, R. E. (1968) Abnormal glucose tolerance in manganese deficient guinea pigs, *J. Nutr.* **94**, 89–94.

45. Flinn, R. H., Neal, P. A. and Fulton, W. B. (1941) Industrial manganese poisoning, *J. Ind. Hyg. Toxicol.* **23**, 374–387.

46. Forth, W. and Rummel, W. (1973) Iron absorption, *Physiol. Rev.* **53**, 724–792.

47. Friedman, S. and Kaufman, S. (1965) 3,4-Dihydroxy-phenylethylamine β-hydroxylase: a copper protein. *J. Biol. Chem.* **240**, 552–554.

48. Friedman, N. and Rasmussen, H. (1970) Calcium, manganese and hepatic gluconeogenesis, *Biochem. Biophys. Acta* **222**, 41–52.

49. Gillette, J. R., Watland, D. and Kalnitsky, G. (1954) The catalysis of the oxidation of some dehydroxybenzene derivatives by various metallic ions. *Biochem. Biophys. Acta* **13**, 526–532.

50. Gillette, J. R., Watland, D. and Kalnitsky, G. (1955) Some properties of the manganese and copper catalyzed oxidation of catechol and some other ortho-dihydroxybenzene derivatives, *Biochim. Biophys. Acta* **16**, 51–57.

51. Goodwin, F. K. and Sack, R. L. (1973) Affective disorders: the catecholamine hypothesis revised In: E. Usdin and S. H. Synder, (Eds.), *Frontiers in Catecholamine Research*. Pergamon Press, Oxford, pp. 1157–1164.

52. Greenberg, D. M. and Campbell, W. W. (1940) Studies in mineral metabolism with the aid of induced radioactive isotopes-IV Manganese, *Proc. Natn. Acad. Sci.* **26**, 448–450.

53. Gupta, S. K., Murthy, R. C. and Chandra, S. V. (1980) Neuromelanin in manganese-exposed primates, *Toxicol. Lett.* **6**, 17–20.

54. Hill, R. M., Holtkamp, D. E., Buchanan, A. R. and Rutledge, E. K. (1950) Manganese deficiency in rats with relation to ataxia and loss of equilibrium, *J. Nutr.* **41**, 359–371.

55. Hornykiewicz, O. (1972) Dopamine and extrapyramidal motor function and dysfunction, *Res. Publ. Assoc. Res. Nerv. Ment. Dis.* **50**, 390–412.

56. Hurley, L. S., Everson, G. T. and Geiger, J. F. (1958) Manganese deficiency in rats: congenital nature of ataxia, *J. Nutr.* **66**, 309–319.

57. ITRC Memoir No. 1 (1976) Outbreak of paralysis in villages of Unnao, Uttar Pradesh—A preliminary epidemiological study.

58. Jaksch, R. V. (1902) In: *Prager Med. Wochensehr.* **18**, 213–214. Quoted from NAS, Manganese, 1973.

59. Kawamura, R. H., Ikuta, S., Fukuzumi, R., Yamada, S., Tsubaki, T., Kodoma and Kurata, S. (1941) Intoxication by manganese in well water, *Kitasato Arch. Exp.* **18**, 145–169.

60. Keefer, R. C., Barak, A. J. and Boyett, J. D. (1970) Binding of manganese and transferrin in rat serum, *Biochem. Biophys Acta* **221**, 390–393.

61. Kimura, M., Yagi, N. and Itokawa, Y. (1978) Effect of subacute manganese feeding on serotonin metabolism in the rat, *J. Environ. Path. Toxicol.* **2**, 455–461.

62. Kaassen, C. D. (1974) Biliary excretion of manganese in rats, rabbits and dogs, *Toxicol. Appl. Pharmacol.* **29**, 458–468.

63. Leach, R. M., Jr. (1976) Metabolism and function of manganese. In: A. S. Prasad (Ed.), *Trace Elements in Human Health and Disease*, Vol. II, Academic Press, New York, pp. 235–247.

64. Le Feure, M. E. and Joel, D. D. (1977) Intestinal absorption of particulate matter, *Life Sci. Stal.* **28**, 96–97.

65. Lindberg, O., and Ernster, L. (1954) Manganese as a co-factor of oxidative phosphorylation, *Nature* **173**, 1038–1039.

66. Maynard, L. S. and Cotzias, G. C. (1955) The partition of manganese among organs and intracellular organelles of the rat, *J. Biol. Chem.* **214**, 489–495.

67. McGeer, E. G., Gibson, S. and McGeer, P. L. (1967) Some characteristics of brain tyrosine hydroxylase, *Can. J. Biochem.* **45**, 1557–1563.

68. McLeod, B. E. and Robinson, H. F. (1972) *Brit. J. Nutr.* **27**, 229. Quoted from *Metals in the Environment*, H. A. Waldron (Ed.), Academic Press, London pp. 206.

69. Mella, H. (1924) The experimental production of basal ganglia symptomatology in Macacus Rhesus, *Arch. Neurol. Psychiat.* **2**, 405–410.

70. Mena, I., Horiuchi, K., Burke, K. and Cotzias, G. C. (1969) Chronic manganese poisoning. Individual susceptibility and absorption in iron, *Neurology* **19**, 1000–1006.

71. Mena, I., Roman C. and Beca, J. P. (1978) In *Proceedings of the 2nd International Congress of the World Federation of Nuclear Medicine and Biology.*

72. Mena, I. (1980) Manganese. In: M. A. Waldron (Ed.), *Metals in the Environment*, Academic Press, London pp. 199–220.

73. Miller, R. C., Keith, T. B., McCarty, M. A. and Thorp, W. T. S. (1940) Manganese as a possible factor influencing the occurrence of lameness in pigs, *Proc. Soc. Exp. Biol. Med.* **45**, 50–51.

74. Miller, S. T., Cotzias, G. C. and Evert, H. A. (1975) Control of tissue manganese, initial absence and sudden emergence of excretion in the neonatal mouse, *Amer. J. Physiol.* **229**, 1080–1084.

75. Moffat, K., Loe, R. S. and Hoffman, B. M. (1976) The structure of metmanganoglobin, *J. Mol. Biol.* **104**, 669–685.

76. Mustafa, S. J. and Chandra, S. V. (1971) Levels of 5-hydroxytryptamine, dopamine and norepinephrine in whole brain of rabbits in chronic manganese toxicity, *J. Neurochem.* **181**, 931–933.

77. National Academy of Sciences (1973) *Manganese, Medical and biological effects of environmental pollutants*, National Academy of Sciences, Washington, D.C., pp. 4–6, 91, 107, 124, 127.

78. Neff, N. H., Burnett, R. E. and Costa, E. (1969) Selective depletion of caudate nucleus dopamine and serotonin during chronic manganese dioxide administration to squirrel monkeys. *Experientia, 25,* 1140–1141.

79. Niyogi, T. P. (1958) Chronic manganese poisoning, *Indian J. Ind. Med.* 3, 3–13.

80. O'Dell, B. L. and Campbell, B. J. (1970) Trace elements: Metabolism and metabolic function. In: M. Florkin and E. H. Stotz (Eds.) *Metabolism of Vitamins and Trace Elements*, Vol. 21, Elsevier, The Netherlands, pp. 223–229.

81. Papavasiliou, P. S., Miller, S. T. and Cotzias, G. C. (1966) Role of liver in regulating distribution and excretion of manganese, *Amer. J. Physiol.* **211,** 211–216.

82. Papavasiliou, P. S., Miller, S. T. and Cotzias, G. C. (1968) Functional interactions between biogenic amines, 3′-5′-cyclic AMP and manganese, *Nature* **220,** 74–75.

83. Penalver, R. (1957) Diagnosis and treatment of manganese intoxication, Report of a case, *AMA Arch. Ind. Health* 16, 64–66.

84. Pentschew, A., Ebner, F. F. and Kovatch, R. M. (1963) Experimental manganese encephalopathy in monkeys, *J. Neuropath. Exp. Neurol.* **22,** 488–499.

85. Plumlee, M. P., Thrasher, D. M., Beesen, W. M., Andrews, F. N. and Parker, H. E. (1956) The effects of manganese deficiency upon the growth development and reproduction of swine, *J. Anim. Sci.* **15,** 352–367.

86. Prasad, A. S. (1978) Manganese In: *Trace Elements and Iron in Human Metabolism*, Plenum, New York, pp. 191–201.

87. Rawal, M. L. (1968) Manganese poisoning in manganese mines in India, *Ind. J. Ind. Med.* 14, 2, 41–51.

88. Rehnberg, G. L., Hein, J. F., Carter, S. D. and Laskey, J. W. (1980) Chronic manganese oxide administration to preweanling rats: Manganese accumulation and distribution, *J. Toxicol. Env. Hlth.* **6,** 217–226.

89. Rodier, J. (1955) Manganese poisoning in Moroccan miners, *Brit. J. Ind. Med.* **12,** 21–35.

90. Roonwal, G. S. (1981) Manganese nodules in Indian ocean, *Science Reporter* **18,** 384–391.

91. Seth, P. K., Hussain, R., Mushtaq, M. and Chandra, S. V. (1977) Effect of manganese on neonatal rat: Manganese concentration and enzymatic alterations in brain, *Acta Pharmacol. Toxicol.* **40,** 553–560.

92. Seth, P. K., Hong, J. S., Kilts, C. D. and Bondy, S. C. (1981) Alteration of cerebral neurotransmitter receptor function by exposure of rats to manganese, *Toxicol. Lett.* **9,** 247–254.

93. Singh, J., Hussain, R., Tandon, S. K., Seth, P. K. and Chandra, S. V. (1974) Biochemical and histopathological changes in early manganese poisoning, *Environ. Physiol. Biochem.* **4**, 16–23.

94. Singh, S., Shukla, G. S. and Chandra, S. V. (1979) The interaction between ethanol and manganese in rat brain, *Arch. Toxicol.* **41**, 363–365.

95. Sitaramayya, A., Nagar, N. and Chandra, S. V. (1974) Effect of manganese on enzymes in the rat brain, *Acta Pharmacol. Toxicol.* **35**, 185–190.

96. Shils, M. E. and McCollum, E. V. (1943) Further studies on the symptoms of manganese deficiency in the rat and mouse, *J. Nutr.* **26**, 1–19.

97. Stokinger, H. E. (1980) The Metals: In: G. D. Clayton and F. E. Clayton (Eds.), *Patty's Industrial Hygiene and Toxicology*, Vol. 24, Wiley, New York, pp. 1749–1769.

98. Tanaka, A. and Lieben, J. (1969) Manganese poisoning and exposure in Pennsylvania, *Arch. Environ. Health* **19**, 674–684.

99. Tauber, F. W. and Krause, A. C. (1943) The role of iron, copper, zinc and manganese in the metabolism of the ocular tissues with special reference to lens, *Amer. J. Opthalmol.* **26**, 260–266.

100. Thomson, A. B. R., Olatunbosun, D. and Valberg, L. S. (1971) Interrelation of intestinal transport system of manganese and iron, *J. Lab. Clin. Med.* **78**, 642–655.

101. Thomson, A. B. R. and Valberg, L. S. (1972) Intestinal uptake of iron, cobalt and manganese in the iron-deficient rat, *Am. J. Physiol.* **223**, 1327–1329.

102. Tichy, M. and Cikrt, M. (1972) Manganese transfer into the bile in rats, *Arch. Toxicol.* **29**, 51–58.

103. Tolonen, M. (1972) Industrial toxicology of manganese, *Work-Environ. Hlth* **9**, 53–60.

104. Underwood, E. J. (1971) In: *Trace Elements in Human and Animal Nutrition.* E. J. Underwood (Ed.), Academic Press, London, pp. 180.

105. Van Bogaert, L. and Dallemagne, M. J. (1945) Approaches experimentales des troubles nerveux du manganisme, *Psychiat. Neurol.* **111**, 60–89.

106. Von Oettingen, W. F. (1935) Manganese: Its distribution, pharmacology and health hazards, *Physiol. Rev.* **15**, 175–201.

107. Weiner, N. (1970) Regulation of norepinephrine biosynthesis, *Ann. Rev. Pharmac.* **10**, 273–290.

108. Whitlock, C. M., Amuso, S. J. and Bittenbender, J. B. (1968) Chronic neurological disease in two manganese steel workers, *Amer. Industr. Hyg. Assoc. J.* **27**, 454–459.

Chapter 7

Mercury and Abnormal Development of the Fetal Brain

B. H. Choi

1. Introduction

There has been mounting concern about environmental pollutants as possible health hazards in recent years. Particularly alarming is the fact that some of the contaminants are found in the food we eat and in the air we breathe. Almost daily we are besieged with reports of environmental contamination of one kind or another, possibly affecting the lives of thousands of individuals. Although intoxications related to industrial exposures have been known for a long time, the tragic outbreaks of mercury (Hg) poisoning in Japan in the 1950s heightened our awareness of potential human intoxication by contaminants in foodstuffs.

Associated with a steadily increasing rate of industrial, cultural, and technological development and advancement, there has been an enormous increase in the rate of emission of various environmental pollutants during the past century. It has been estimated that more than a thousand new chemical substances are being introduced annually throughout the world. These include

B. H. Choi: Division of Neuropathology, Department of Pathology, University of California, Irvine, California College of Medicine, Irvine, California

pharmaceutical, industrial, agricultural, and laboratory reagents, in addition to a large variety of food and fuel additives. Yet only a fraction of these substances has been adequately tested for their toxicity, carcinogenicity, mutagenicity, or teratogenicity. Almost any biologically active substance can act as a pollutant.

Any Hg found in humans and animals probably represents a pollutant from the environment because there is no known metabolic requirement for Hg or Hg-compounds in living organisms. Mercury intoxication resulting from industrial exposure or following medicinal use has been known for centuries. The sources and uses of Hg[125], its toxicity and metabolism, and the ecological implications of Hg in the environment have been the subject of many excellent reviews.[1,18,36,55,74,84,122]

There are various chemical forms of Hg that can cause toxic effects in humans and animals. However, from the standpoint of environmental health hazards, methylmercury (MeHg) is the most important. It has been demonstrated that methylation of inorganic Hg occurs in the sediment of waterways[130] and in the gut through the action of microorganisms; therefore, any Hg that enters water or the gut is a potential source of MeHg. Furthermore, it is also known that Hg found in fish is practically all in the form of MeHg.[114] MeHg is particularly dangerous since it is the form most toxic to humans,[78] since it affects the central nervous system (CNS) and since it remains in the body for protracted periods of time.[55]

The possibility of fetal Hg intoxication was considered many years ago when a high frequency of abortions was observed among women undergoing treatment for syphilis,[2] and when Hg was detected in the tissues of stillborn infants of mothers treated with Hg-compounds.[95] Subsequent experiences in Japan[56,57] and Iraq[5-7] have clearly demonstrated that severe MeHg intoxication involving pregnant mothers may be lethal to the fetus, while lesser degrees of intoxication may lead to some form of cerebral palsy accompanied by mental retardation. There have been reports[5,6,14,44,50,79,90-92,126,131] indicating that MeHg in experimental animals and in humans readily passes from mother to fetus and that, once across the placenta, it may have a greater affinity for the fetal than the adult CNS. It is particularly noteworthy that, following severe *in utero* poisoning, an infant may appear completely normal at birth, only to develop psychomotor deficits as the nervous system matures.[117,118] The seriousness of the problem is underlined by observations in humans that indicate that brain levels of MeHg may be higher in the fetus than the mother, and that the developing nervous system *in utero* may be much more vulnerable to the toxic effects of MeHg than the adult nervous system, a difference in susceptibility that has already been documented in experimental animals.[50,67,96] Thus, although a gravid woman might sustain only minor CNS damage from MeHg intoxication, the consequences to the fetus could be much more serious.

Of most immediate concern are the effects of long-term exposure to Hg and Hg-compounds, particularly those following consumption of Hg-contaminated food by pregnant women. Obviously the risk of this type of long-term exposure also involves the population in general. In some parts of the world MeHg has become a serious socioeconomic and public health problem, and fetal brain damage contributes significantly to the gravity of that problem.[85]

The sensitivity of the human fetal CNS to the toxic effects of MeHg *in utero* is now well-established. The clinical manifestations resulting therefrom include various cerebral palsy syndromes encompassing mild to marked psychomotor disorders and mental retardation. However the pathological basis for the neurological derangements in affected children remains largely unknown.

Injury to the fetal brain by MeHg might be expected to result in developmental deviations that are related to major neuroembryological events occurring at the time of exposure. The developing human brain undergoes a programmed series of developmental events during ontogenesis. These events include proliferation and migration of neurons and other cell types, selective aggregation, programmed cell death, elongation of axons, and establishment of synapses, differentiation, and maturation, and so on. Organogenesis is complete at 3 months, and a full quotient of nerve cells is attained by the time of birth or beyond, after which there is no further numerical increase. The weight of the human brain increases quite rapidly during fetal life, and somewhat more slowly during the first 2 yr postnatally. The total mass of the brain attains a maximum by the twelfth year, although certain parts of the brain continue to grow up to the fifth decade. It is apparent, therefore, that the entire period of brain development covers not only *in utero* growth, but a lengthy postnatal period as well.

An understanding of the nature of developmental deviations of the CNS must be based upon a thorough knowledge of the normal pattern of growth and development in the fetal brain. Equally important is a basic knowledge of the principles of teratology.[129] Conclusions regarding the teratogenicity of physical, microbiological, or chemical agents must be drawn with care. The physiolo-

gical factors and pathogenetic mechanisms involved in teratogenesis are, after all, complex and poorly understood. Susceptibility to teratogenesis varies not only with the genetic make-up of
the conceptus, but also with the developmental stage at the time
of exposure to the agent in question.[129]

With the application of investigative tools such as
radioautography, electron microscopy (EM), tissue culture techniques, and so on, significant progress has been made in the field
of developmental neurobiology in recent years. Thus, a considerable amount of information is now available for certain laboratory
animals concerning the time of origin and the pattern of migration
of neurons, and the differentiation and maturation of various parts
of the CNS.[3,4,11,12,60,102]

It is said that approximately 75% of all fetal deaths and about
40% of deaths during the first year of life may be the result of CNS
malformation in humans. The estimation is that nearly a third of
all embryos do not survive beyond the second month because of
such malformations.

Although inheritance of certain congenital anomalies is well-
established, not all neural malformations are genetically determined. Even among those that are, not all arise from germ cell defects transmitted from one generation to the next. It is quite
possible for a situation to arise in which normal germ cells produce a normal zygote, and in which a subsequent genetic muta-
tion occurs before the developmental program has proceeded very
far. It is said that the series of genetically preprogrammed instruc-
tions carried in the structure of DNA in human differentiation
would be equivalent to 10 million printed pages if a single letter of
the alphabet were defined by six bits. Yet, remarkably, errors in the
developmental program occur in less than 7% of all human
births.[72] Some believe that most, if not all, cancers have environmental causes[40] and that the same circumstances might apply to
birth defects.[72] But even today, among those malformations that
can be attributed to genetic transmission and chromosomal aberrations, which comprise approximately 25% of all birth defects,
only about 10% can be ascribed to environmental factors.

The difficulty in identifying the cause(s) of CNS malformations rests not only on the observation that most anomalies may
result from the potentiative effects of multiple factors, but also
that the same structural anomalies may be caused by genetic as
well as exogenous causes. Furthermore, elucidation of the pattern of normal human CNS growth and development has of necessity depended largely on the use of static morphological stud-

ies of autopsy or surgical material. Although much has been learned of the times of origin, sites of proliferation, migration, and differentiation of developing CNS cells from analysis of material at defined time intervals, our knowledge of human CNS development has not progressed very far from the days of Ramón y Cajal (1890).[107] A recent resurgence of the use of Golgi methods, the development of immunohistochemical methods using specific chemical markers[9,10,16,41,59,112] and the use of EM and other tools, such as morphometric devices and computers, have provided some basic data concerning human CNS development.[25,26,30,32,33,54,103,104,113]

In this chapter the prenatal effects of MeHg upon fetal brain development, particularly in the human, will be described primarily from the standpoint of pathologic anatomy.

I was fortunate enough to have had an opportunity to be exposed to the teachings of the late Dr. Harry Goldblatt in Cleveland, Ohio. When Dr. Goldblatt was awarded the gold-headed cane for his monumental contributions to the field of pathology, he said that, during his continued endeavors in experimental pathology, he always found pathologic anatomy to be an essential and important crutch. In this context, it should also be recalled that it was the careful and thorough analysis of autopsied material by Professor Takeuchi and his coworkers[121] in the Department of Pathology at Kumamoto University in Japan that provided the critical clues that permitted the identification of the then mysterious illness called "Minamata disease" as being caused by MeHg poisoning.

2. Abnormal Development of Human Fetal Brain From Methylmercury Poisoning In Utero

Although occasional clinical reports of human fetal poisoning by Hg have appeared in recent years,[13,42,56,57,116] the only descriptions of which this author is aware that deal with fetal CNS pathology following maternal ingestion of MeHg are those of Matsumoto et al.[82] and Choi et al.[35]

In the Japanese cases, Matsumoto et al.[82] emphasized that the pathologic lesions within the CNS of prenatally intoxicated children were "very similar in type and distribution to these found in adult patients suffering from Minamata disease . . ."; however, in both of their cases they noted evidence of malformation of the brain, which was characterized by the "remains of matrix cells in

the region of ventricular walls, an abnormal architectural arrangement of cortical neurons, poorly myelinated areas and hypoplasia
of corpus callosum." They also described the diffuse nature of the
pathologic changes compared to the focal selective nature of pathologic lesions found in adult cases.

The cases that we have described were derived from a cata-
strophic outbreak of MeHg poisoning associated with consumption of home-made bread prepared from the seed grain of wheat
that had been treated with fungicide containing MeHg. This outbreak, which occurred in Iraq during the winter of 1971–1972, affected more than 6500 people and resulted in 459 deaths. Roughly
half of the patients were female, a substantial number of whom
were of child-bearing age. Many aspects of this outbreak have been
analyzed,[5–7,13,51] and long-term followup studies are still in
progress.[80]

2.1. Clinical Summaries of the Iraqi Cases

The first of the two cases in our report[35] was born to a 28-yr-old
mother who had consumed MeHg-contaminated bread three or
four times daily for approximately 10 weeks, starting from the sixth
to eighth week of gestation. The infant, a female, was born at term
after an uncomplicated labor and delivery. Birth weight was 2890
g. The child, who exhibited no dysmorphic features, appeared
healthy initially. Because of an episode of fever and diarrhea she
received large doses of antibiotics with what was judged to be a
good response. She died unexpectedly on the 33rd day of interstitial pneumonitis, which was confirmed at autopsy. The blood levels of Hg during life ranged from 442 to 658 ng/mL.

Early in pregnancy, the mother had experienced malaise, severe abdominal pain, blurring of vision, and numbness of the
tongue, perioral region, hands, and feet. A few weeks later she developed shoulder and knee joint pains. When admitted to the hospital at 6 months' gestation, she was noted to have moderately
slurred speech, decreased visual acuity, decreased peripheral visual fields, diminished sensation of the hands and the feet, and exaggerated deep tendon reflexes in all four extremities. Following
discharge from the hospital her symptoms and signs gradually improved. Except for decreased peripheral visual fields and moderately increased deep tendon reflexes in all four extremities, she appeared relatively normal when examined 2 yr later.

The second infant, also female, was born at term to a 20-yr-old
woman who had consumed MeHg-contaminated bread three to

four times daily for a period of approximately 10 weeks, starting from the eighth to tenth week of gestation. Labor and delivery were uncomplicated, and the birth weight was 3178 g. No dysmorphic features were noted. The Apgar score was 8. Grasp, Moro, and sucking reflexes were within normal limits. The neurological examination was reported to be normal. Because of the high Hg content of the blood at birth (1568 ng/mL), an exchange transfusion was carried out within 4 h of delivery. She was exchanged with 450 cc of O-negative blood. During the procedure she developed pallor and bradycardia, requiring resuscitation. Despite these measures, death occurred at approximately 7 h, shortly after completion of the exchange transfusion.

The clinical features of our two cases differ somewhat from those reported by Matsumoto et al.[82] The mothers of both of our cases exhibited signs and symptoms of severe MeHg intoxication, and the infants, though seemingly normal at birth, died during early postnatal life. In the Japanese cases, on the other hand, the mothers were asymptomatic, and the infants, who lived for 6 yr, 3 months and 2 yr 6 months, respectively, suffered from severe cerebral palsy associated with repeated convulsive seizures.

2.1.1. NEUROPATHOLOGIC FINDINGS

2.1.1.1. Gross Pathology.
Both brains were smaller than normal, and the estimated weight of each, extrapolated from the weight of the material available for pathologic study, was 250 g. (The normal weight of a neonatal human brain is approximately 330 g.) The external topography of each was characterized by irregularities of gyral size and configuration and by incomplete opercularization (Fig. 1). In coronal section, except for a few heterotopic islands of grey matter and a reduction in the volume of cerebral white matter, no other macroscopic abnormalities were noted.

2.1.1.2. Histopathology.
Microscopic alterations in the two brains were remarkably similar except for minor differences in the severity of involvement. The principal findings consisted of the presence of large numbers of heterotopic neurons in cerebral and cerebellar white matter and disturbances of cerebral and cerebellar cortical neuronal alignment and organization.

Heterotopic neurons, which occurred both singly and in groups, exhibited varying degrees of differentiation. Many were large pyramidal-shaped neurons (Fig. 2b), but groups of small, dark, rounded or elongated immature neurons were also noted, especially in the deep white matter (Fig. 2d). In the cerebrum scat-

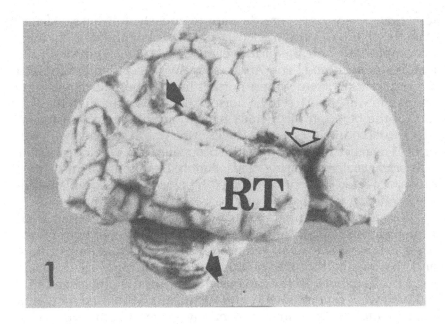

Fig. 1. Gross photograph of the brain of Case 2. Note simplified gyral pattern. Frontal lobe is short and opercularization is incomplete. The superior temporal gyrus is narrower than the mid-temporal gyrus. RT: Right temporal lobe. Thick arrows point to leptomeningeal melanosis. (by permission of *J. Neuropathol. Exp. Neurol.* 37, 723, 1978).

tered foci of heterotopic grey matter that contained both neurons and glial cells were identified within the leptomeninges (Fig. 2a). There was also patchy melanosis of the leptomeninges (Fig. 1).

The laminar pattern of the cerebral cortex was disturbed and in many places exhibited irregular undulations (Figs. 2c, 3a). Irregular grouping or vertical alignment of neurons (Fig. 3b) and admixtures of large and small neurons were very common. In many areas the neurons were often not aligned perpendicular to the pial surface, their apical dendrites pointing upward at various angles to the perpendicular. The deeper cortical layers were not distinctly defined in many areas and gradually blended into the white matter. The deeper nuclear structures of the basal ganglia and the thalamus were within normal limits. Cellular and tissue architecture within the hippocampus were normal.

Groups of small immature neurons were noted in the subependymal regions of the basal ganglia and thalamus as well as

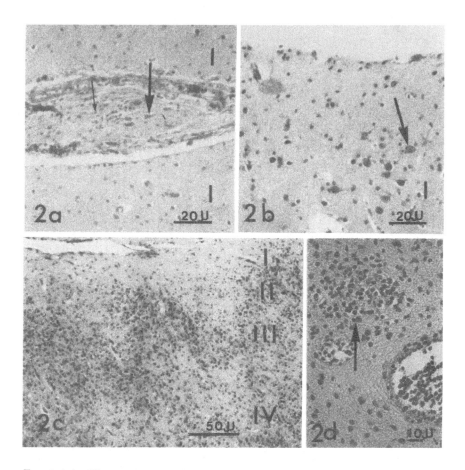

FIG. 2. (a) Photomicrograph of a heterotopic island of grey matter in the subarchnoid space containing neurons (large arrow) and glial cells (smaller arrow): I, layer I of the cerebral cortex (Case 1). (b)

FIG. 2. (b) Photomicrograph of a group of large pyramidal neurons (arrow) in layer I (labeled I) of cerebral cortex. This was noted in all regions of the cerebral cortex in both cases (Case 1). (c)

FIG. 2. (c) Photomicrograph demonstrating a disorganized laminar pattern of the cerebral cortex. Note irregular clumping and wavy arrangement of neurons interspersed with relatively less cellular zones: I–IV, cortical layers. Superior frontal gyrus of Case 2. (d)

FIG. 2. (d) Photomicrograph of heterotopic islands of small immature neurons (arrow) in the deep white matter of the frontal lobe (Case I) (by permission of J. Neuropathol. Exp. Neurol. 37, 725, 1978, Figs. 2a–2d).

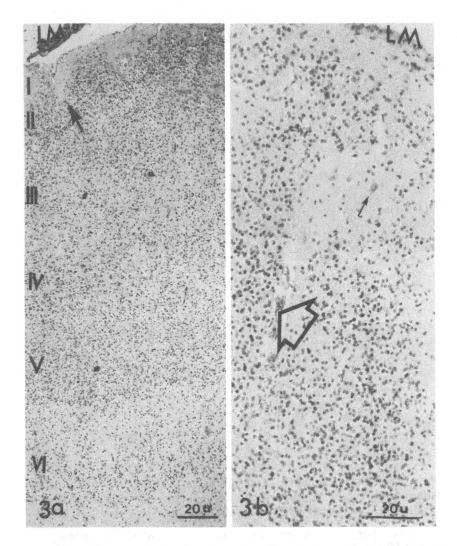

Fig. 3. (a) Photomicrograph showing irregular undulation (arrow) of cerebral cortical neurons: I–VI, cortical layers of cerebrum; LM, leptomeninges.

Fig. 3. (b) Photomicrograph showing columnar arrangement (empty arrow) and abnormal clumping of cortical neurons of insular cortex of cerebrum. Large pyramidal neurons (small arrow) are scattered among smaller granular neurons.

deep within the white matter of the internal capsule and basal ganglia, particularly around blood vessels and within myelinating long fiber tracts.

The general architectural pattern of the cerebellum was within normal limits. Large numbers of heterotopic neurons in the folial and medullary white matter were noted throughout (Figs. 4a, 4b, 5). Although there were differences in the degree of development of different folia (as one would expect in normal development), there was no specific abnormality of the layering pattern of the cerebellum in either case. The thickness and pattern of formation of both the external and internal granular layers varied from place to place.

An additional prominent feature in the white matter of both cerebrum and cerebellum was the presence of numerous gemistocytic astrocytes containing large amounts of eosinophilic cytoplasm and irregularly shaped nuclei (Fig. 6). Fibers arising from these astrocytes were readily demonstrated by hematoxylin and eosin staining as well as by phosphotungstic acid–hematoxylin staining for glial fibrils. These astrocytes were not associated with either reactive pleomorphic microglia or macrophages. Axons in the white matter were plentiful, as visualized in Bodian preparations. On the other hand, glial elements in the white matter that could be identified as myelin-forming glia or as mature oligodendroglial cells were much less numerous than astrocytes except in a few cerebellar folia in case 1, where myelin formation was relatively well-advanced.

In summary, the principal histopathological findings in the brains of the two infants were widespread neuronal heterotopias in the cerebral and cerebellar white matter and abnormal patterns of organization and distorted alignment of cortical neurons. The additional findings of diffuse astrogliosis without activation of pleomorphic microglia or macrophages and the lack of necrotic foci are worthy of further comment (vide infra).

The nature of the histopathologic changes observed in these two newborn infants represents the outcome of disturbances in brain development, more specifically, abnormal neuronal migration and derangement in the fundamental structuring of grey matter. Heterotopia and heterotaxia of neurons and disordered cortical organization are, to be sure, seen in a variety o' conditions in man[37] and in a number of experimental disorders.[75,124127] They are therefore nonspecific changes. There is, however, strong presumptive evidence for implicating MeHg as the causal agent:

1. The blood levels of Hg in both infants at birth were markedly elevated.

2. The postmortem brain levels of Hg were much above those reported in normal infants in Minamata, Japan.[121]

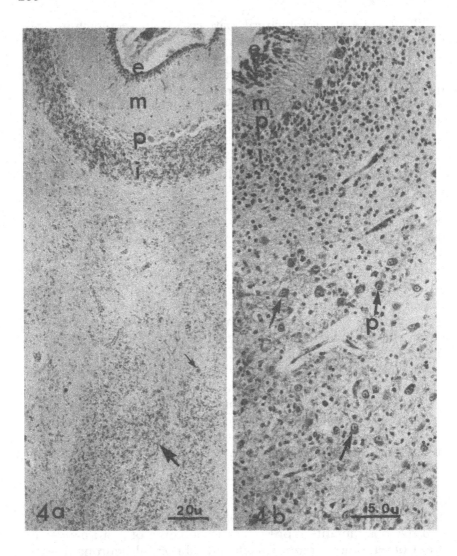

FIG. 4. (a) Photomicrograph showing groups of heterotopic neurons composed of small granular neurons (large arrow) and scattered large pyramidal neurons (smaller arrow) in the deep white matter of cerebellum (Case 1): e, external granular layer; m, molecular layer; p, Purkinje cell layer; i, internal granular layer. (b) Photomicrograph showing many heterotopic Purkinje cells in the deep white matter of cerebellum (Case 2). Note the well-differentiated pyramidal-shaped neurons (p) and scattered smaller neurons. The internal granular layer is not well formed: e, external granular layer; m, molecular layer; p, Purkinje cell layer; i, internal granular layer.

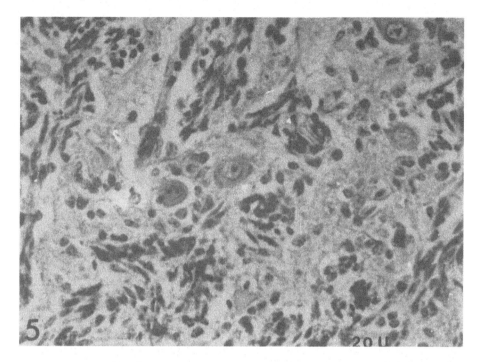

Fig. 5. Photomicrograph showing a group of large pyramidal shaped neurons and small round to elongated neurons in the white matter of the cerebellum. Case 2.

3. The exposures to MeHg in both infants occurred during the critical period of neuronal migration, which normally begins at about 7 weeks' gestation and continues into the third trimester.[113] This contention is based on a reliable clinical history from both mothers concerning the time of onset and duration of exposure to MeHg as a consequence of MeHg-contaminated bread ingestion.

4. The blood levels of Hg in both mothers when hospitalized during pregnancy were at toxic levels for adults. The Hg-levels in hair segments in one of the mothers provided an excellent opportunity to document the rising levels of Hg dating from the start of ingestion of MeHg-contaminated bread, thereby providing relative blood values for both mother and infant.

5. The striking similarity of the pathologic anatomy of the brains of our two infants to those of Matsumoto et

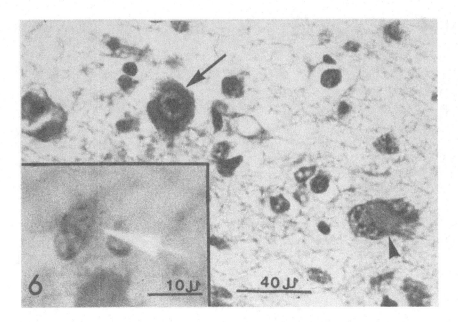

Fig. 6. Photomicrograph showing heterotopic neurons (arrow) and gemistocytic astrocyte (arrow head) in the white matter of the cerebrum. Case 2. Inset depicts an astrocyte with granules of mercury (white arrow) in the cytoplasm, as seen by photoemulsion histochemistry. Case 2. (by permission of *J. Neuropathol. Exp. Neurol.* **37**, 727, 1978).

al.[82] suggests a common pathogenetic link to MeHg poisoning.

An important negative finding in our cases was the absence of focal neuronal destructive changes in the cerebrum and cerebellum, such as that which has been described in children and adults, or in infants exposed postnatally to MeHg.

The large numbers of astroglia in the white matter must be accepted as indicative of a reactive phenomenon to some type of destructive process. However, the nature of this white matter insult, its time of occurrence, and its significance remain obscure. Although there were definitely fewer numbers of myelin-forming glia and mature forms of oligodendrocytes than astrocytes in the cerebral and cerebellar white matter in both cases, and although myelin had begun to develop in cerebellar folia in the older infant, neither frank necrosis nor macrophages were present in either case. In addition, axons in the white matter were not remarkable. It is also noteworthy that the astrocytes had the same generalized cerebral and cerebellar white matter distribution as the widespread neuronal heterotopias, and that many of them contained Hg grains, as demonstrated by photoemulsion histochemistry[81] (Fig. 6). With these features in mind, we believe that it is possible that MeHg might have caused or at least contributed to the production of the astrocytic response.

Taking into account all the histopathological and toxicological findings in our cases, the possibility that the principal manifestations of disturbed neuronal migration and the exuberant white matter astrocytic reaction were caused by MeHg merits serious consideration. Thus the disturbances in the normal growth and development of neurons and astroglia, particularly the intimate neuronal–glial interaction that exists during neuronal migration,[31,101] might well represent the fundamental mechanisms underlying the complex pathological changes encountered in these infants.

3. Experimental Studies on the Effects of Prenatal Methylmercury Poisoning Upon Fetal Brain Development

3.1. Experiments Using Animal Models

There have been numerous experimental studies dealing with the effects of MeHg poisoning *in utero* upon fetal growth and development using a variety of animal species. These have included frogs,[38] Japanese medakas,[39] killifish,[128] mallard eggs,[61] chicks,[110] mice,[47,48] rats,[70] hamsters,[58] and cats.[68,69] Most of these studies, however, have been directed primarily toward the study of reproductive, embryocidal, and gross teratogenic effects and have not necessarily been concerned with morphological changes in the developing CNS. A few examples will be cited. Gale and Ferm[49] administered 7.5, 8, and 10 mg/kg of mercuric acetate and phenyl-mercuric acetate to pregnant golden hamsters on the 8th day (d) of gestation. A small number of fetuses showed miscellaneous CNS anomalies such as exencephaly, anencephaly, and anophthalmia. Similarly, Harris et al.[58] reported a high incidence of fetal resorption, a decrease in fetal weight and moderate to high frequency of malformations, among surviving newborn hamsters, after injection of 8 mg/kg of methylmercuric chloride (MMC) on the 5th, 8th, and 9th d of pregnancy. Murakami[90,91] reported studies in mice and rats following the administration of MMC at 0–12 d gestation, and described the development of hydrocephalus, focal hemorrhage, and edema, along with "severe damage in internal capsule, caudate and putamen and abundant pyknosis of cerebellum." Fuyuta et al.[47] observed decreased body weight, fetal death, and an increasing incidence of malformations and cleft palate in mice and rats following oral administration of MMC to pregnant animals. Based on this study Fujimoto et al.[43] demonstrated the protective effects of Tiopronin (2-mercaptopropionyl glycine) against cleft palate formation after MMC poisoning in pregnant animals. Su and Okita[119,120] also reported the occurrence of cleft palate among the offspring of mice born to mothers treated with MMC. Olson and Massaro[97,98] observed delayed closure of the palate and reduced levels of cAMP and phosphodiesterase in the offspring of mice treated with 10 mg/kg of MeHg, but postulated that the action of MeHg on palate closure was not a direct effect of MeHg on components of the adenyl cyclase system. Mottet[88] was unable to detect any gross or microscopic changes in the CNS of rats prenatally poisoned with methyl mercury hydroxide, but observed a dose-dependent rise in the rate of maternal deaths, neonatal resorption, and fetal deaths along with a reduction in fetal weight. Chen et al.[22] likewise observed no significant histopatho logical lesions in the CNS, but noted weight reduction of prenatally intoxicated rat fetuses as the only significant finding. They thought this phenomenon to be associated with deceased DNA synthesis in MeHg-exposed fetuses. Spyker and Smithberg[117] found that a single injection of methyl mercury dicyanamide given to inbred strains of pregnant mice caused developmental defects that varied with the day of treatment, the strain, and the dose administered. Thus there is ample evidence to show that MeHg is embryotoxic and teratogenic in a variety of animal species. Unfortunately however, most of these teratological or toxicological studies have not placed sufficient emphasis upon the pathologic anatomy to permit an understanding of the effects of MeHg compounds upon fetal CNS development.

On the other hand, Spyker et al.[118] have reported subtle behavioral abnormalities in the offspring of mice born to mothers who were treated with MeHg compounds during pregnancy. The newborn animals initially appeared grossly normal, but subsequently showed significant and measurable abnormalities in various behavioral tests, indicating delayed functional and physiological effects of MeHg poisoning *in utero.* Since that time there have been numerous other reports of behavioral abnormalities in the offspring of animals prenatally poisoned with MeHg.[17,21,62,89,99,110] The literature contains only a handful of reports dealing with the histopathologic changes in the fetal or newborn nervous systems of animals prenatally poisoned with MeHg. Morikawa[87] produced changes in the fetal cat cerebellum by maternal administration of bis-ethyl mercuric sulfide. The changes described include smallness of the cerebellum with marked loss of granule cells and relative preservation of Purkinje cells. Brief note was also made of heterotopic localization of Purkinje cells. Matsumoto et al.[83] described both gross and microscopic abnormalities in the cerebellum of fetal rats following maternal administration of single doses of 2–20 mg/kg of MMC on gestational days 9, 10, and 11. Their description and illustrations, however, are not sufficiently detailed to permit adequate assessment of the changes observed. Nonaka[94] examined newborn rat brains by light and EM after oral administration to pregnant rats of 2 mg/kg of MMC daily for 20 d. Disappearance of nucleoli and reduced numbers of chromatin granules and ribosomes in some of the granule cells of the cerebellum were described. Purkinje cells were also noted to be either electron-dense or vacuolated. Khera[68] observed reduced populations of external granule cells and focal aggregates of heterotopic cells in cat cerebella after prenatal intoxication with MeHg. More recently Chang and his coworkers[19-21] have described light and EM changes in the cerebellum of mice poisoned with MeHg *in utero*. Their findings included "accumulation of electron-dense bodies, presumably lysosomes, in both Purkinje neurons and granule cells. . . . disorientation of rough endoplasmic reticulum and the presence of tubular structures and dilatation of smooth endoplasmic reticulum." In addition, "segmental atrophy of axon and thinning of myelin sheaths in some nerve fibers and abnormal synaptic complexes, displaying a reduction or lack of postsynaptic densities" were described. The significance of these changes and their relationship to the behavioral changes observed in some of these animals are not clear at the present time.

The descriptions thus far reported of the pathologic anatomy of the CNS following *in utero* exposure to Hg and Hg-compounds must be considered too nonspecific and lacking in detail to allow an adequate understanding of the neuropathology of fetal MeHg poisoning on brain development.

A world of caution must be expressed at this time. The structure of the brain is more complex than that of any other organ and undergoes profound changes during its protracted ontogenesis. Thus the degree of individual variability at certain periods of growth and development must be well-understood and known before proper assessment of any alleged deviation can be made. In particular, subtle changes that have been described using various histopathological techniques are often esoteric and unreliable in unskilled hands and may be too subjective and limited in scope. Frequently what has been described as abnormal may simply be nonspecific or artifactual, and may not bear any relationship to the functional and neurological derangements or behavioral abnormalities observed. In fact, in some instances the effects may be too subtle for morphological delineation using currently available techniques.

More recently, our laboratory has carried out a number of experiments designed to provide insight into the effects of MeHg upon specific regions of the developing mouse CNS. Cerebellar development in mice is primarily a postnatal phenomenon. The cerebellum has a specific architecture composed of specific cell types, and its normal growth and development have been studied extensively. Moreover, the cerebellum has been a favored site of pathological changes following MeHg poisoning in humans as well as in experimental animals.

Newborn mice (C57BL/6J) were injected subcutaneously once a day on postnatal days 3, 4, and 5 with 5.0 mg/kg body weight of MMC labeled with ^{203}Hg (totaling 15 mg/kg body weight in each animal). Control animals received physiological saline in place of MMC. The animals were sacrificed on postnatal day 15. Perfusion with Karnovsky's fixative was carried out and representative sections of cerebellum and other parts of CNS were prepared for light microscopic, Golgi, radioautographic, histochemical, and EM studies. Whole body counts of radioactivity were made daily with a gamma counter before and after injection of MMC until the day of sacrifice. Tritiated thymidine was also injected in each animal to assess the pattern of DNA synthesis of cells undergoing active proliferation.

A significant finding in this study was the absence of clearance of radioactive Hg from the body during the period of experimentation. The clearance half-times calculated by least-square linear regression of daily whole body counts after correction for radioactive decay remained constant at the end of the 15th postnatal day. The absence of clearance of ^{203}Hg in neonatal mice has also been noted by other investigators and represents an as yet unexplained metabolic phenomenon. There was considerable uptake of ^{203}Hg-labeled MeHg in the brains of these neonatal animals, however. Important negative findings in this study included an absence of grossly detectable changes in the cerebellum and the lack of any apparent histological abnormality either in 6 μm paraffin-embedded sections or in 1.0-μm epon-embedded sections. Using Golgi methods and EM, however, two significant findings were noted. The first was demonstrated by the rapid Golgi method and consisted of greatly simplified dendritic arborization of many Purkinje cells in MeHg-treated animals, even though the overall number of Purkinje cells and the general architectonic pattern of the cerebellar cortex were maintained. Extensive survey of electron micrographs revealed marked swelling of the cytoplasm of pericapillary astroglia. Associated with this was an attenuation of cytoplasm of endothelial cells.[29]

The mechanism by which MeHg leads to the incomplete development and differentiation of Purkinje cell dendrites remains unknown. Further studies using larger doses and/or longer periods of exposure to MeHg may bring about more apparent changes in the cerebellum; however, this preliminary study indicates the importance of studying specific developmental features using specific procedures in correlation with other morphological and functional parameters. The changes in the astroglia and vascular endothelium are of a regressive nature and may not bear any relation to dendritic differentiation but may indicate an important role for astroglia and the vascular system in overall cerebellar development.

In another study a single intraperitoneal injection of MMC labeled with [203]Hg was made on day 16 of gestation. Control animals received physiological saline. Animals were allowed to come to term, and the offspring were sacrificed on postnatal days 5, 10, and 20. Whole body counts of radioactivity in the mothers during the period of gestation and during lactation following delivery were made daily in a gamma counter. One day prior to the time of sacrifice, 1.0 μCi/mL of tritiated thymidine was injected intraperitoneally into the offspring (both control and experimental).

The clearance half-time of radioactivity in lactating mothers was 8.7 d, indicating a relative increase of clearance in lactating mothers compared to controls (10–14 d).[51]

Relatively high radioactivity was retained in the brains of MMC-treated offspring. A statistically significant reduction of body weight in MeHg-treated litters was noted. Preliminary Golgi surveys of the cerebellum showed marked differences in the degree of differentiation of Purkinje cells (Fig. 7) between MeHg-treated and control animals. Radioautographic analysis of tritiated thymidine uptake in the external granule cells of the vermis showed an apparent delay in the proliferative activity of the MeHg-treated group.

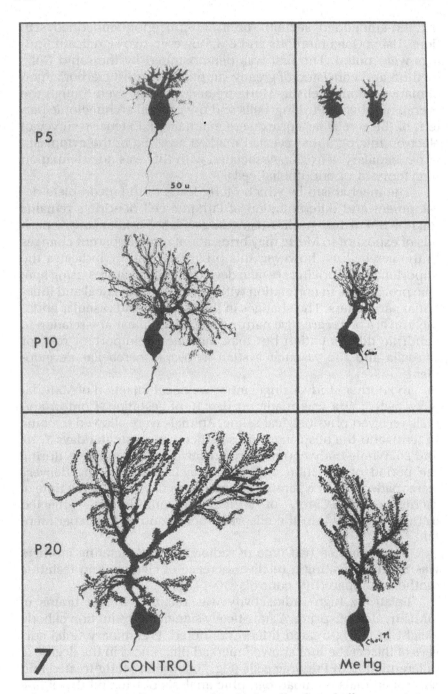

FIG. 7.   Camera lucida tracings of Purkinje cells of cerebellum on postnatal days 5 (p5), 10 (p10), and 20 (p20) of control group and methyl mercury-treated group. Rapid Golgi preparation. MeHg was injected into mothers on day 16 of gestation, 15 mg/kg).

Among MeHg-intoxicated animals the labeling index was 9.3% on postnatal day 5 (control 8.6%) and 18.1% on postnatal day 10 (control 6.2%), indicating a progressive decrease of proliferative activity in control group while the MeHg group was still undergoing active incorporation of thymidine, presumably signifying continuation of proliferative activity prior to migration of external granule cells into the internal granular layer.

The results of this study, which indicate a delay in dendritic arborization of Purkinje cells and a delay in the proliferative activity of external granule cells following *in utero* MeHg intoxication, clearly illustrate the value of a correlative morphological approach in bringing out some of the more subtle effects of MeHg upon the developing CNS.

3.2. Experiments Using Human Fetal Brain Cells In Vitro

To test the hypothesis that MeHg causes abnormal neuronal migration, we have carried out a number of experiments using organotypic and monolayer cultures of human fetal brain cells obtained from aborted fetuses. Our studies, as well as those of others, have demonstrated that many of the developmental events that occur in normal fetal brain are repeated in cultures of fetal brain tissue.

In order to determine the characteristic pattern of cell migration under normal conditions in vitro, repeated observations were made using time-lapse cinematography, phase and electron microscopy, and immunohistochemical techniques. Migration of neurons occurs within 3 d of explantation of human fetal cerebral cortical fragments. Neuronal migration under these circumstances is an orderly process that follows a similar sequence regardless of the mode of culture (organotypic, dissociated, primary or secondary) or the type of surface upon which the cells are grown (plastic or glass, with or without collagen or other supportive substrates).

Extension of neurites and astrocytic processes occurs at the margin of explants within 48–72 h. This is followed by the outward migration of astrocytes. Many small dark neurons follow in rapid succession. Astrocytes continue to proliferate and migrate and soon cover the peripheral portions of the explants. The pulsating tips of neurites anchor themselves to the processes or cytoplasm of astrocytes and to other neurites as well. While thus securely fastened, translocation of nerve cell somata is effected by movement of their nuclei into the extended neuritic processes and by subsequent withdrawal toward the cell body of the trailing neurites. Neurons or groups of neurons frequently follow the same pathway, using astroglial processes as guides. This sequence of events is repeated until large numbers of neurons have reached the outgrowth area. A remarkable phenomenon is the intimate association between astrocytes and neuronal cell bodies and their neurites. Neurons tend to cluster along astrocytic processes during migration, and groups of neurons may then be carried away to more distant locations as the astrocytes move about, particularly when organotypic cultures are subjected to trypsinization or simple mechanical dissociation. Continuous observation of organotypic cultures using time-lapse cinematography clearly demonstrates that there is active in vitro migration of human fetal neurons, and that the sequence of events is relatively orderly.[24,28]

There are several viewpoints regarding the mode of neuronal migration in the developing vertebrate CNS. According to the so-called karyokinesis theory,[15] the nuclei of postmitotic germinal cells within the ventricular zone move outward along cylindrical cytoplasmic extensions. Another theory proposes that the post-mitotic neurons migrate along glial processes.[33,101] Our study demonstrates that both of these phenomena take place in culture. Although neurons, either singly or in groups, tend to use astroglial cells (cell bodies and their processes) as guides for movement to distant locations, translocation of neuronal somata is accomplished by nuclear movement within extended neuritic cylinders. The intracytoplasmic movement of neuronal nuclei can occur in either direction within bipolar neurons.

In a study of this kind, particularly when dealing with cultured cells, it is vital to identify precisely the types of cells being dealt with. We have previously emphasized the need for using combined techniques of cell identification, correlating the features observed by phase microscopy of living cells, optical microscopy of stained preparations, immunocytochemistry using specific cell markers,[9-11] and EM. The isolation and subsequent production of antibodies to glial fibrillary acidic (GFA) protein by Eng[41] and others[16,25] has been a major breakthrough in the identification of astroglial cells in vivo and in vitro.[9,10]

Using all of the techniques outlined above, as well as radio-autography, we have demonstrated a special type of junctional complex between neurites and astrocytic membranes in culture (Fig. 8), but the functional significance of this type of junctional complex is not clear at this time. However, repeated observations of the intimate association between neurons and astrocytes both in vivo and in vitro,[24,31,32] coupled with the known beneficial effects of astrocytes upon the survival of immature neurons in cul-

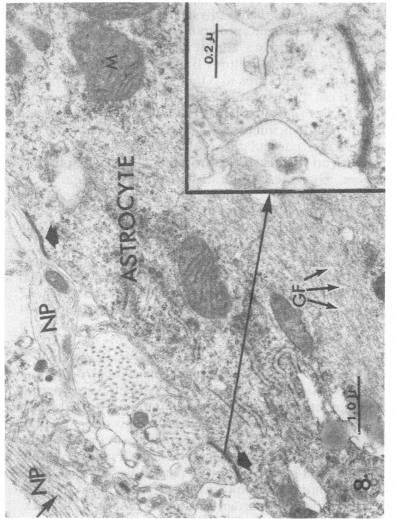

Fig. 8. Electron micrograph demonstrating the formation of intermembranous electron dense junctions (thick arrows) between neuronal processes (NP) and astrocytic membrane. GF: glial filaments.

ture,[31] strongly support the concept that the interaction of migrating neurons with astrocytes is important to the ontogenesis of the developing CNS.

To test the effects of MeHg upon neuronal migration under these experimental conditions, explants of human fetal brain tissue in complete culture media were exposed to various concentrations of MMC.[28] In control dishes physiological saline rather than MMC was added to the culture medium.

Addition of MMC causes abrupt cessation of rather active movement of cells in these cultures under time-lapse cinematographic observation. Exposures to 0.1 and 0.2 mM resulted in the rapid development of cytotoxic damage to thin neurites and neuronal cell bodies as well as to astrocytes. The damage was severe and immediate. The initial site of damage appeared to be the neuritic membrane, particularly in the vicinity of the growth cones.[28] By electron microscopy, disruption and degeneration of neurites were characterized initially by disappearance of neurotubules. Thus it would appear that the MMC-exposed neurons failed to extend their neurites not only because of membrane damage, but also because of the lack of neurotubules necessary for structural support and for axoplasmic transport.

The cytotoxic actions of MMC could be stopped immediately if the cultures were washed and placed in fresh normal media. Provided that the toxic effects were not too far advanced, surviving neurons and astrocytes recovered slowly and resumed normal cell motility, including migratory activity, 48–72 h following removal of the MMC.

Damage occurred most rapidly and most severely to thin neurites and neuronal cell bodies, while damage to astrocytes developed somewhat later, although both types of cells were irreversibly affected after 10 min in 0.2 mM or 20 min in 0.1 mM MMC.[28] Following exposure to much lower concentrations (e.g., 0.01 mM) of MMC for 30–40 min, the progression of damage to the astrocytic membrane could be halted by placing the cells in normal culture media, after which a slow recovery to normal was observed.

Meso-2,3-dimercaptosuccinic acid (DMSA) is known to be effective in decreasing body burden and brain content of Hg in experimental animals. This compound has also been found to be much less toxic than agents such as BAL (British anti-Lewisite) or penicillamine.

Experiments with secondary monolayer cultures of human fetal astrocytes demonstrated that the addition of 0.01 mM MMC caused rapid retraction and coagulative cell membrane changes under time-lapse cinematography. After 30–40 min of MMC exposure, cultures were washed and replaced with complete media containing 1.0 mM DMSA. The damaged cells reformed smooth membrane with active filopodial motion and ruffling. Exposures to DMSA alone in cultures over a 12-h period did not cause damage, at least on phase-microscopic observation. DMSA and MMC in combination likewise prevented cellular damage under similar experimental conditions.[34]

Further studies of MMC-exposed astroglial cells by EM demonstrated almost complete disappearance of neurotubules from the matrix of neurites as well as astroglial processes. The mitotic astroglial cells in MMC-exposed cultures showed a complete lack of mitotic spindles in addition to various other cytotoxic changes. DMSA treatment restored polymerization of neurotubules in these cells within a short period of time.[34]

Although the results of tissue culture experiments cannot be interpreted literally as phenomena that are occurring in the living organism, the system allows in-depth analysis of molecular events taking place between living human fetal brain cells and MMC under precisely controlled conditions. It is obvious that the specificity of MeHg action on human fetal brain cells should be further investigated by comparing the effects of other chemicals known to produce specific toxic effects on the homeostasis of neurotubules (such as colchicine),[73] microfilaments (cytochalasin B), and other cytoskeletal elements.

Many questions remain unanswered. What are the mechanisms involved in the damaging and reparative phenomena seen in these experiments? How is the damaged membrane put together again to form a new membrane with the addition of the complexing agent? What is the pattern of MeHg distribution before and after the addition of complexing agents? These are some of the questions that must be addressed in future investigations.

The pattern of DNA synthesis within human fetal astrocytes was also investigated using autoradiographic techniques. The results showed that levels above 3.99 μM MMC profoundly inhibited DNA synthesis, whereas at levels below 1.19 μm there was no demonstrable inhibition.[27]

Since astrocytes play a major role in neurogenesis, the effects of MeHg on astroglia must be taken into account when evaluating the prenatal effects of MeHg upon CNS growth and development.

Tissue culture systems, of course, have inherent limitations. They also offer certain advantages. The features of particular value are the precise control of the environment, the ease of application of many research techniques, and the direct accessibility of cells for observing developmental phenomena within living cells.

There have been many reports on tissue culture studies of the effects of MeHg. Kim[71] observed severe degeneration of nerve cells, particularly granule cells and Purkinje cells, in myelinating cultures of newborn mouse cerebellum after exposure to 0.4–10.0 mM methyl-mercury acetate. Kasuya[65,66] conducted a series of experiments using cultures of chick and rat dorsal root ganglia to test the effects of various Hg-compounds by determining the degree of outgrowth of cells and fibers as an indicator of cytotoxicity. He postulated that the cytotoxic effects of MeHg resulted from the binding of Hg-compounds to hydrophobic and hydrophilic moieties at specific membrane sites, and further tested modification of such actions by vitamin E and selenium. Ammitzboll and Clausen[8] demonstrated the cytotoxic activity of 0.1 mM MMC upon chick embryo brain cells as well as the reversibility of such toxicity following removal of the MMC and replacement with normal culture media.

Other tissue culture studies of MeHg effects have utilized a variety of cell types, including mouse leukemic cells,[93] mouse glioma cells,[86,100] HeLa S3 cells,[52,53] and mouse neuroblastoma cells.[73,100] Although there are certain merits to using neoplastic cells for the study of MeHg toxicity, the relevance of these models to the effects of MeHg upon fetal brain development is open to question.

One must also be aware of the dangers of extrapolating to humans results that have been obtained in rodent experiments, although information thus obtained clearly may provide important clues to guide future research in an area such as MeHg toxicity. The use of cultures of human and mouse fetal brain tissue will not only provide data that are more directly applicable to the human fetus, but will also permit direct comparison of changes in mouse and human fetal brain caused by MeHg. As a prelude to defining human (and, therefore, environmental) tolerance levels to MeHg, the systems described promise to provide data that will permit a better understanding of the pathogenesis of the toxic effects of MeHg upon the growing fetal brain.

4. Comment

Upon analysis of the many experimental and relatively few human studies of MeHg intoxication *in utero*, it is obvious that our knowledge of the nature of the cellular and tissue abnormalities responsible for clinical neurological dysfunction in MeHg-affected children is extremely limited. Even less is known about the developmental or morphological events taking place in the fetus itself at the time of MeHg exposure or the mode of pathogenesis of abnormal fetal brain development. Many factors undoubtedly contribute to fetal wastage, growth retardation, mental retardation, or functional disturbances that become manifest later in life. In spite of numerous experimental studies that have been designed to elucidate the prenatal effects of MeHg poisoning and to establish the neurotoxic nature of MeHg poisoning, a large gap still exists in our understanding of the nature of the abnormalities in fetal brain development and of the manner of clinical presentation of MeHg toxicity.

Since the first report of psychomotor retardation in children suspected to have been poisoned by MeHg *in utero*[42] in Sweden, ample clinical evidence has been developed to indicate the vulnerability of the human fetal CNS to the toxic effects of MeHg. However, there are no features that are sufficiently specific to permit the diagnosis of MeHg intoxication in children on the basis of clinical examination alone. The most frequently cited abnormalities include delayed developmental milestones, microcephaly with associated mental retardation, cerebellar symptoms, dysarthria, and primitive reflexes. A few examples of minor deformities of limbs, strabismus, blindness, and deafness have also been reported. In some, these original symptoms gradually improved or worsened as the children grew. In many there was no abormality initially, yet symptoms of neurological dysfunction became manifest later in life. Many unsuccessful attempts have been made in the past to correlate the concentration of Hg in hair, blood, amniotic fluid, or umbilical cords of affected children with the severity of symptoms.[56]

Recently, Marsh et al.[80] after many years of careful followup studies of 29 gestationally intoxicated mother–infant pairs of fetal MeHg poisoning in Iraq demonstrated for the first time a relationship between maximal maternal hair Hg concentrations and the frequency of neurological abnormalities in affected children. They have shown that maternal hair concentrations in the range of 99–384 parts per million (ppm) are associated with a significantly higher frequency of abnormal neurological findings than those in two groups having lower maternal hair Hg concentrations (12–85 and 0–11 ppm). Although most of the mothers experienced some transient to severe signs of Hg intoxication, in some no symptoms were seen during pregnancy.

The most critical prenatal interval for the development of neuropathologic changes following exposure to MeHg appears to be the late embryonic and fetal period, during which neuronal migration and histogenetic events such as differentiation and organization of cerebral and cerebellar cortical plates are taking place, although the precise mechanisms involved remain largely speculative. There is less susceptibility to lethal injury or to gross organ deviation at this time than at an earlier phase of embryonic development. Adverse influences during neuronal migration and organization of cortical development in contrast tend to produce more subtle histologic abnormalities that may become clinically manifest as psychomotor retardation or other neurological disturbances only at a later date, i.e., postnatally.

This is in contradistinction to the situation in MeHg-intoxicated children or adults, in which neurons within specific regions are affected more or less selectively.[63] The effects upon neurons during earlier phases of development may not be readily apparent on examination of neonatal brains because of the difficulty in identifying the sites of nerve cell damage or loss in the absence of vigorous cellular (i.e., macrophage or astrocyte) response.

As with other teratogenic agents, MeHg has been assumed to have relatively imperceptible effects prior to the blastocyst stage. In general, exposures to toxic agents during the period of gametogenesis results in death of the sperm or egg, chromosomal defects or embryonic death leading to early abortion. However, there have been reports of MeHg-induced polyploidy in the root cells of plants,[105] *XXY* nondisjunction in the fruit fly (*Drosophilia melanogaster*),[106] and chromosome breakage in human lympocytes.[115] All of these data indicate a need for further investigation. At the molecular level, interaction of MeHg with microtubular proteins may be of fundamental importance in the development of CNS abnormalities. We have shown that MMC exerts profound effects upon the neurotubules of human fetal neurons and astrocytes in culture.[28] Others have also described the effects of MeHg upon mitotic spindle fibers in certain types of neural tumors.[86]

For a better understanding of the effects of MeHg upon the developing CNS, however, more work is needed to clarify the *normal* sequence of neuroembryological events, both at the molecular and cellular levels. For example, we have shown that in the human cerebrum completion of DNA synthesis within neurons destined for the fronto-parietal cortex occurs much earlier than previously believed.[30] At the cellular level, it has been suggested that germinal cells give rise first to neurons and subsequently to glial cells.[45,46] Our studies and those of others,[25,76,77] however, strongly suggest the likelihood that neurogenesis and gliogenesis take place concomitantly in developing CNS. Until these and other uncertainties regarding neurogenesis are resolved, therefore, it will be difficult to ascertain precisely the molecular mechanisms and stages of development that are applicable in prenatal MeHg intoxication. Furthermore, the same pathogenetic mechanisms may not be operative at all stages of gestation. Cell death may be minimally important at one stage because of the ability to replace damaged or dead cells. At another stage, however, cell death may be of great importance because of the loss of such ability, resulting in permanent cell depletion in critical regions of the brain. Other possibilities exist. The teratogenic effects may appear immediately and result in cell death or embryonic death, or they may appear much later and be ascertainable only in the postpartum or adult period.

Subtle neurophysiological and behavioral abnormalities in the offspring of prenatally injured mothers are the most difficult to evaluate. This is not only because of difficulties in differentiating the effects of other environmental factors, but also because there has been practically no attempt to correlate such findings with neuroanatomical changes except for a few recent attempts by some.[108,109]

It should be realized that the brain is not a homogenous organ such as the liver or the thyroid. It is composed of many different organs of differing structure and function. Therefore, a description such as degeneration of cerebrum or cerebellum cannot suffice to characterize the abnormalities involved. Descriptions should be more specific about the structures, cell types, organelles, and stage of development involved. For example, the program of development differs for each folium of the cerebellar vermis.[64] Therefore, specific vermal lobules should be indicated when statements are made regarding focal abnormalities, such as thickness in the external granule layer. Reference should be made to the precise timing of myelin formation within the region in question wherever statements are made regarding MeHg-induced abnormalities and myelination.

Techniques are now available to isolate different cellular and subcellular organelles within the developing CNS.[23,132] Specific effects of MeHg upon cellular and molecular organelles isolated during neurogenesis should be individually analyzed prior to drawing conclusions regarding the effects of MeHg damage upon the fetal brain.

It should also be realized that the developing CNS has only a limited capacity to react to injury. Infiltration by inflammatory cells, for example, is not seen until 6 months of fetal age. The residue of cell and organelle degeneration such as are known to occur during normal embryrogenesis may not be apparent on histologic examination of the neonatal brain. A neuronal loss of 25% or more may be necessary before it can be appreciated with conventional histological methods. Residual gliosis is often the only evidence of prior tissue or cell damage.

For a proper study of the problem, intercommunication among scientists of many different disciplines is essential. To a toxicologist, an electron-dense Purkinje cell soma may seem ominous; yet it may merely represent an artifact of tissue preparation or an isolated nonspecific finding. Along these lines, speculative attempts to explain observed behavioral abnormalities in experimental animals on the basis of electron-dense cerebellar internal granule cells or synaptic abnormalities may be inappropriate. For these reasons, it is highly desirable to enlist the cooperative efforts of the activities of developmental neurobiologists, pathologists, neurophysiologists, toxicologists, teratologists, epidemiologists, and behavioral psychologists, as well as neurologists, to delineate the neurotoxic effects of MeHg, particularly on the developing fetal brain. Only careful correlative observation and recording of subtle changes of behavior, pathologic anatomy, or pattern of clinical abnormality will lead to meaningful results.

Along with experiments in suitable animal models, it should be recognized that the molecular and subcellular interaction between MeHg and developing human fetal brain cells can be more easily and economically studied in vitro. The events associated with neuronal migration, interaction of neurons and glia, formation of transient or permanent cellular junctions, development of specific molecules in specific cell types, branching patterns of dendrites, and so on, can be isolated and studied using this system. It is also essential to develop criteria for the recognition of MeHg intoxication at different stages of gestation.

This should be achieved not only for acutely toxic states, but also for long term low-level exposures. The effects upon various enzyme systems, membranes, cell types, and subcellular organelles, must all be studied.

A retrospective study of MeHg intoxication in Niigata, Japan, showed that there was only one case of prenatal MeHg poisoning.[123] Pregnant women with more than 50 ppm of Hg in their hair were advised to terminate their pregnancies during that outbreak.[56] Whether or not this was a reliable method of determining critical threshold levels of fetal toxicity needs to be evaluated further. The example of Seveso, Italy is a case in point. Following elective abortion in 35 pregnant women who were exposed to toxic fumes only one fetus showed demonstrable anomalies. On the other hand, there is no way of knowing whether or not subtle chronic effects of prenatal intoxication might have become manifest in their offspring had their pregnancies not been aborted. Retrospective and prospective studies should be carried out in pregnant women exposed to Hg and Hg-compounds to determine whether there is any rise in the frequency of abortions, stillbirths, congenital defects, cerebral palsy, mental retardation, and subtle behavioral abnormalities. Most of all, the coordinated efforts of scientists of different disciplines are needed to provide a better estimate of the limits of safety for the fetus following maternal MeHg exposure.

Acknowledgments

The author wishes to express his deep appreciation for help given by Drs. Ronald C. Kim, Lowell W. Lapham, and Thomas Clarkson. He also wishes to express thanks to the kind permission granted by the *J. Neuropathol. Exp. Neurol.* (editor-in-chief: Dr. John Moossy) for presentation of cases published in that Journal. The excellent secretarial help of Ms. Lucia Wisdom and technical help of Ms. Teresa Espinosa is greatly appreciated. (Supported in parts by NIEHS Grants RO 1 ES 02928 and ES 01247.)

References

1. Aarsonson, T. (1971) Mercury in the environment. *Environment* 13, 16–23.
2. Afonso, J. and DeAlvarez, R. (1960) Effects of mercury on human gestation. *Am. J. Obstet. Gynecol.* 80, 145–154.
3. Altman, J. (1972) Postnatal development of the cerebellar cortex in the rat. II. Phases in the maturation of Purkinje cells and of the molecular layer. *J. Comp. Neurol.* 145, 349–464.

4. Altman, J. and Anderson, W. J. (1972) Experimental reorganization of the cerebellar cortex. I. Morphological effects of elimination of all microneurons with prolonged x-irradiation started at birth. *J. Comp. Neurol.* **146**, 355–406.

5. Amin-Zaki, L., Elhassani, S., Majeed, M. A., Clarkson, T. W., Doherty, R. A. and Greenwood, B. S. (1974) Intra-uterine methylmercury poisoning in Iraq. *Pediat.* **54**, 587–595.

6. Amin-Zaki, L., Elhassani, S., Majeed, M. A., Clarkson, T. W., Doherty, R. A., Greenwood, B. S. and Giovanoli-Jakubezak, T. (1976) Perinatal methylmercury poisoning in Iraq. *Am. J. Dis. Child.* **130**, 1070–1076.

7. Amin-Zaki, L., Elhassani, S., Majeed, M. A., Clarkson, T. W. and Greenwood, B. S. (1978) Methylmercury poisoning in Iraqi children: clinical observations over two years. *Brit. Med. J.* **1**, 597–666.

8. Ammitzboll, T. and Clausen, J. (1973) The toxic effect of methylmercury chloride on brain cell cultures from chick embryo. *Environ. Physiol. Biochem.* **3**, 248–254.

9. Antanitus, D. S., Choi, B. H. and Lapham, L. W. (1975) Immunofluorescence staining of astrocytes *in vitro* using antiserum to glial fibrillary acidic protein. *Brain Res.* **89**, 363–367.

10. Antanitus, D. S., Choi, B. H. and Lapham, L. W. (1976) The demonstration of glial fibrillary acidic protein in the cerebrum of the fetus by indirect immunofluorescence. *Brain Res.* **103**, 613–616.

11. Angevine, Jr., J. B. (1970) Time of neuron origin in the diencephalon of the mouse. An autoradiographic study. *J. Comp. Neurol.* **139**, 129–188.

12. Angevine, J. B. and Sidman, R. L. (1961) Autoradiographic study of cell migration during histogenesis of cerebral cortex in the mouse. *Nature* **192**, 766–768.

13. Bakir, F., Damluji, S. F., Amin-Zaki, L., Murtadha, M., Khalidi, A., Al-Rawi, N. Y., Takriti, S., Chahir, H. I., Clarkson, T. W., Smith, J. C. and Doherty, R. A. (1973) Methylmercury poisoning in Iraq. *Science* **181**, 230–240.

14. Berlin, M. and Ulberg, S. (1963) Accumulation and retention of mercury in the mouse brain: a comparison of exposure to mercury vapor and intravenous injection of mercuric salt. *Arch. Environ. Health* **12**, 33–42.

15. Berry, M. and Rogers, A. W. (1965) The migration of neuroblasts in the developing cerebral cortex. *J. Anat.* **99**, 691–709.

16. Bignami, A., Eng, L. F., Dahl, D. and Uyeda, C. T. (1972) Localization of glial fibrillary acidic protein in astrocytes by immunofluorescence. *Brain Res.* **43**, 429–435.

17. Bornhausen, M., Müsch, H. R. and Greim, H. (1980) Operant behavior performance changes in rats after prenatal methylmercury exposure. *Tox. Appl. Pharmacol.* **56**, 305–310.

18. Brown, J. R. and Kulkarni, M. V. (1967) A review of the toxicity and metabolism of mercury and its compounds. *Med. Services J. Canada* **23**, 786–808.

19. Chang, L. W. and Reuhl, K. (1977) Ultrastructural study of the latent effects of methyl mercury on the nervous system after prenatal exposure. *Environ. Res.* **13**, 171–185.

20. Chang, L. W., Reul, K. R. and Lee, G. W. (1977) Degenerative changes in the developing nervous system as a result of in utero exposure to methylmercury. *Environ. Res.* **14**, 414–423.

21. Chang, L. W., Reuhl, K. and Spyker, J. M. (1977) Ultrastructural study of the latent effects of methylmercury on the nervous system after prenatal exposure. *Environ. Res.* **13**, 171–185.

22. Chen, W., Body, R. L. and Mottet, K. N. K. (1979) Some effects of continuous low-dose congenital exposure to methylmercury on organ growth in the rat fetus. *Teratology* **20**, 31–36.

23. Cheung, M. and Verity, M. A. (1981) Methylmercury inhibition of synaptosome protein synthesis. Role of mitochondrial dysfunction. *Environ. Res.* **24**, 286–298.

24. Choi, B. H. (1979) Mechanism of neuronal migration in human foetal cerebrum in vitro. *Yonsei Med. J.* **20**, 92–104.

25. Choi, B. H. (1981) Radial glia of developing human fetal spinal cord: Golgi, immunohistochemical and electron microscopic study. *Dev. Brain Res.* **1**, 249–267.

26. Choi, B. H. (1981) Hematogenous cells in developing human embryos and fetuses. *J. Comp. Neurol.* **196**, 683–694.

27. Choi, B. H., Cho, K. H. and Lapham, L. W. (1981) Effects of methylmercury on DNA synthesis of human fetal astrocytes in vitro. *Brain Res.* **202**, 238–242.

28. Choi, B. H., Cho, K. H. and Lapham, L. W. (1981) Effects of methylmercury on human fetal neurons and astrocytes *in vitro:* A time-lapse cinematographic, phase and electron microscopic study. *Environ. Res.* **24**, 61–74.

29. Choi, B. H., Kudo, M. and Lapham, L. W. (1981) A Golgi and electron microscopic study of cerebellum in methylmercury-poisoned neonatal mice. *Acta Neuropathol.* (Berl.) **54**, 233–237.

30. Choi, B. H. and Lapham, L. W. (1974) Autoradiographic studies of migrating neurons and astrocytes of human fetal cerebral cortex *in vitro. Exp. Mol. Pathol.* **21**, 204–217.

31. Choi, B. H. and Lapham, L. W. (1976) Interactions of neurons and astrocytes during growth and development of human fetal brain *in vitro. Exp. Mol. Pathol.* **24**, 110–125.

32. Choi, B. H. and Lapham, L. W. (1978) Radial glia in the human fetal cerebrum: a combined Golgi, immunofluorescent and electron microscopic study. *Brain Res.* **148**, 295–311.

33. Choi, B. H. and Lapham, L. W. (1980) Evolution of Bergmann glia in human fetal cerebellum: A Golgi, immunofluorescent and electron microscopic study. *Brain Res.* **190**, 369–383.

34. Choi, B. H. and Lapham, L. W. (1981) Effects of meso-2,3-dimercaptosuccinic acid on methylmercury injured human fetal astrocytes in vitro. *Exp. Mol. Pathol.* **34**, 25–33.

35. Choi, B. H., Lapham, L. W., Amin-Zaki, L. and Saleem, T. (1978) Abnormal neuronal migration, deranged cerebral cortical organization, and diffuse white matter astrocytosis of human fetal brain: A major effect of methylmercury poisoning in utero. *J. Neuropathol. Exp. Neurol.* **37**, 719–733.

36. Clarkson, T. W. (1972) Recent advances in the toxicology of mercury with emphasis on the alkylmercurials. *Crit. Rev. Toxicol.* **1**, 203–234.

37. Crome, L. (1952) Microgyria. *J. Pathol. Bact.* **64**, 479–495.

38. Dial, N. A. (1976) Methylmercury: Teratogenic and lethal effects in frog embryos. *Teratology* **13**, 327–334.

39. Dial, N. (1978) Methylmercury: Some effects on embryogenesis in the Japanese Medaka, Oryzias Latipes. *Teratology* **17**, 83–92.

40. Doll, R. (1977) Strategy for detection of cancer hazards to man. *Nature* (Lond) **265**, 589–596.

41. Eng, L. T. (1980) The glial fibrillary acidic (GFA) protein. In: R. Bradshaw and D. Schneider, (Eds.), *Proteins of the Nervous System*, Raven Press, New York, pp. 85–117.

42. Engleson, G. and Herner, T. (1952) Alkylmercury poisoning. *Acta Paediatr. Scand.* **41**, 289–294.

43. Fujimoto, T., Fuyuta, M., Kiyofuji, E. and Hirata, S. (1979) Prevention by Tiopronin (2-mercaptopropinoyl glycine) of methylmercuric chloride-induced teratogenic and fetotoxic effects in mice. *Teratology* **20**, 297–302.

44. Fujita, E. (1969) Experimental studies of organic mercury poisoning. The behavior of the Minamata disease causing agent in maternal bodies, and its transfer to their infants via either placenta or breast milk. *J. Kumamoto Med. Soc.* **43**, 47.

45. Fujita, S. (1963) The matrix cell and cytogenesis in the developing central nervous system. *J. Comp. Neurol.* **120**, 37–42.

46. Fujita, S. (1967) Quantitative analysis of cell proliferation and differentiation in the cortex of the postnatal mouse cerebellum. *J. Cell. Biol.* **32**, 277–287.

47. Fuyuta, M., Fujimoto, T. and Hirata, S. (1978) Embryotoxic effects of methylmercuric chloride administered to mice and rats during organogenesis. *Teratology* **18**, 353–366.

48. Fuyuta, M., Fujimoto, T. and Kiyofuji, E. (1979) Teratogenic effects of a single oral administration of methylmercuric chloride in mice. *Acta Anat.* **104**, 356–362.

49. Gale, T. F. and Ferm, V. H. (1971) Embryopathic effects of mercuric salts. *Life Sci.* **10**, Part II, 1341–1347.

50. Garrett, N. E., Burriss, J., Garrett, B. and Archdeacon, J. W. (1972) Placental transmission of mercury to the fetal rat. *Tox. Appl. Pharmacol.* **22**, 649–654.

51. Greenwood, M. R., Clarkson, T. W., Doherty, R. A., Gates, A. H., Amin-Zaki, L., Elhassani, S. and Majeed, M. A. (1978) Blood clearance half-times in lactating and nonlactating members of a population exposed to methyl-mercury. *Environ. Res.* **16**, 48–54.

52. Gruenwedel, D. W., and Cruikshank, M. K. (1979) The influence of sodium selenite on the viability and intracellular synthetic activity (DNA, RNA, and protein synthesis) of HeLa S3 cells. *Tox. Appl. Pharmacol.* **50,** 1–7.

53. Gruenwedel, D. W., and Friend, D. (1980) Long-term effects of methylmercury (II) on the viability of HeLa S3 cells. *Bull. Environ. Contam. Toxicol.* **25,** 441–447.

54. Hamburger, V. (1980) S. Ramon y Cajal, R. G. Harrison, and the beginnings of neuroembryology. *Perspect. Biol. Medicine* **23,** 600–616.

55. Hammond, A. L. (1971) Mercury in the environment: Natural and human factors. *Science* **171,** 788–789.

56. Harada, M. (1976) Minamata disease. Chronology and medial report. *Bull. Inst. Constitutional Medicine,* Kumamoto University **25,** Supp. 1–60.

57. Harada, Y. (1968) Congenital (or fetal) Minamata disease. In: *Minamata Disease. Study Group of Minamata Disease.* Kumamoto Univ., Japan, 93–117.

58. Harris, S. B., Wilson, J. G. and Printz, R. H. (1972) Embryotoxicity of methylmercuric chloride in golden hamsters. *Teratology* **6,** 139–142.

59. Hartman, B. K., Cimino, M., Moore, W. and Agarwal, H. C. (1977) Immunohistochemical localization of brain specific proteins during development. *Trans. Am. Soc. Neurochem.* **8,** 66.

60. Hicks, S. P. and D'Amato, J. (1969) Cell migration to the isocortex in the rat. *Anat. Rec.* **160,** 619–634.

61. Hoffman, D. J. and Moore, J. M. (1979) Teratogenic effects of external egg applications of methylmercury in the Mallard, Anas platyrhynchos. *Teratology* **20,** 453–462.

62. Hughes, J. A. and Sparber, S. B. (1978) D-Amphetamine unmasks postnatal consequences of exposure to methylmercury in utero: methods for studying behavioral teratogenesis. *Pharmacol. Biochem. Behavior* **8,** 365–375.

63. Hunter, D. and Russell, D. S. (1954) Focal cerebral and cerebellar atrophy in a human subject due to organic mercury compounds. *J. Neurol. Neurosurg. Psychiat.* **17,** 235–241.

64. Jacobson, M. (1978) *Developmental Neurobiology.* Plenum Press. New York.

65. Kasuya, M. (1972) Effects of inorganic, aryl, alkyl and other mercury compounds on the outgrowth of cells and fibers from dorsal root ganglia in tissue culture. *Tox. Appl. Pharmacol.* **23,** 136–146.

66. Kasuya, M. (1975) The effects of Vitamin E on the toxicity of alkylmercurials on nervous tissue in culture. *Tox. Appl. Pharmacol.* **32,** 347–354.

67. Kelman, B. J., Steinmetz, S. E., Walter, B. K. and Sasser, B. (1980) Absorption of methylmercury by the fetal guinea pig during mid to late gestation. *Teratology* **21,** 161–165.

68. Khera, K. S. (1973) Teratogenic effects of methylmercury in the cat: Note on the use of this species as a model for teratogenicity studies. *Teratology* **8,** 293–304.

69. Khera, K. S., Iverson, F., Hierlihy, L., Tanner, R. and Trivett, G. (1974) Toxicity of methylmercury in neonatal cats. *Teratology* 10, 69–76.

70. Khera, K. S. and Tabacover, S. A. (1973) Effects of methylmercuric chloride on the progeny of mice and rats treated before or during gestation. *Fd. Cosmet. Toxicol.* 11, 245–254.

71. Kim, S. U. (1971) Neurotoxic effects of alkyl mercury compound on myelinating cultures of mouse cerebellum. *Exp. Neurol.* 32, 237–246.

72. Klingberg, M. A. and Papier, C. M. (1979) Environmental teratogens. Their significance and epidemiologic methods of detection. *Contr. Epidem. Biostatist.* 1, 1–22.

73. Koerker, R. (1980) The cytotoxicity of methylmercuric hydroxide and colchicine in cultured mouse neuroblastoma cells. *Tox. Appl. Pharmacol.* 53, 458–469.

74. Koos, B. J. and Longo, L. D. (1976) Mercury toxicity in the pregnant woman, fetus, and newborn infant. *Am. J. Obstet. Gynecol.* 126, 390–409.

75. Langman, J. and Shimada, M. (1971) Cerebral cortex of the mouse after prenatal chemical insult. *Am. J. Anat.* 132, 355–374.

76. Levitt, P., Cooper, M. L. and Rakic, P. (1981) Coexistence of neuronal and glial precursor cells in the cerebral ventricular zone of the fetal monkey: an ultrastructural immunoperoxidase analysis. *J. Neurosci.* 1, 27–39.

77. Levitt, P. and Rakic, P. (1980) Immunoperoxidase localization of glial fibrillary acidic protein in radial glial cells and astrocytes of the developing rhesus monkey brain. *J. Comp. Neurol.* 193, 417–448.

78. Lu, F. C., Berteau, P. E. and Clegg, D. J. (1972) The toxicity of mercury in man and animals. In: *Mercury Contamination in Man and his Environment*, International Atomic Energy Agency, Vienna, pp. 67–85.

79. Mansour, M. M., Dyer, N. C., Hoffman, L. H., Davies, J. and Brill, A. B. (1974) Placental transfer of mercuric nitrate and methyl mercury in the rat. *Am. J. Obst. Gynecol.* 119, 557–562.

80. Marsh, D. O., Myers, G. J., Clarkson, T. W., Amin-Zaki, L., Tikriti, S., and Majeed, M. A. (1980) Fetal methylmercury poisoning: clinical and toxicological data on 29 cases. *Ann. Neurol.* 7, 348–353.

81. Matsumoto, H., Kameda, T. and Takeuchi, T. (1969) Pathological studies on organic mercury poisoning: supplement to histochemical demonstration of mercury in the brain of Minamata disease. *Adv. Neurol. Sci.* 13, 270–278.

82. Matsumoto, H., Koya, G. and Takeuchi, T. (1965) Fetal minamata disease. *J. Neuropathol. Exp. Neurol.* 24, 563–574.

83. Matsumoto, H., Suzuki, A. and Morita, C. (1967) Preventive effect of pencillamine on the brain defect of fetal rat poisoned transplacentally with methyl mercury. *Life Sci.* 6, 2321–2326.

84. McGregor, J. T. and Clarkson, T. W. (1974) Distribution, tissue binding and toxicity of mercurials. In: Mendel Friedman (Ed.), *Advances in Experimental Medicine and Biology.* Vol. 48., Plenum Press, New York, pp. 463–503.

85. Methylmercury Study Group (1980) McGill Methyl Mercury Study. A study of the effects of exposure to methylmercury on the health of individuals living in certain areas of Province of Quebec. McGill Univ., Montreal, Canada.

86. Miura, K., Suzuki, K. and Imura, N. (1978) Effects of methylmercury on mitotic mouse glioma cells. *Environ. Res.* **17,** 453–471.

87. Morikawa, N. (1961) Pathological studies on organic mercury poisoning. *Kumamoto Med. J.* **14,** 87–93.

88. Mottet, N. K. (1974) Effects of chronic low-dose exposure of rat fetuses to methylmercury hydroxide. *Teratology* **10,** 173–190.

89. Müsch, H. R., Bornhausen, M., Kriegel, H. and Greim, H. (1978) Methylmercury chloride induces learning deficits in prenatally treated rats. *Arch. Tox.* **40,** 103–108.

90. Murakami, U. (1969) Toxicity of organic mercury compounds in prefetus embryo. *Jpn. Med. Assoc.* **61,** 1059–1073.

91. Murakami, U. (1972) The effect of organic mercury on intrauterine life. *Acta Exp. Med. Biol.* **27,** 301–336.

92. Nakamura, K. and Saeki, S. (1967) Preventive effect of penicillamine on the brain defect of fetal rat poisoned transplacentally with metylmercury. *Life Sci.* **6,** 2321–2326.

93. Nakazawa, N., Makino, F. and Okada, S. (1975) Acute effects of mercuric compounds on cultured mammalian cells. *Biochem. Pharmacol.* **24,** 489–493.

94. Nonaka, I. (1969) An electron microscopical study on the experimental congenital minamata disease in rat. *Kumamoto Med. J.* **22,** 27–39.

95. Nordberg, G., and Skerfving, S. (1971) Metabolism. *In* Freiberg, L. and Vostal, J. (eds.), *Mercury in the Environment. A Toxicological and Epidemidological Appraisal.* PB-205000 National Technical Information Service, US Dept. of Commerce, pp. 4–1 to 4–143.

96. Null, D. H., Gartside, R. S., and Wei, E. (1968) Methylmercury accumulation in brains of pregnant, non-pregnant and fetal rats. *Life Sci.* **12,** 65–72.

97. Olson, F. C. and Massaro, E. J. (1977) Effects of methyl mercury on murine fetal amino acid uptake, protein synthesis and palate closure. *Teratology* **16,** 187–194.

98. Olson, F. C. and Massaro, E. B. (1980) Developmental pattern of cAMP, adenyl cyclase, and cAMP phosphodiesterase in the palate, lung, and liver of the fetal mouse: alterations resulting from exposure to methylmercury at levels inhibiting palate closure. *Teratology* **22,** 155–166.

99. Olson, K. and Bousch, G. M. (1975) Decreased learning capacity in rats exposed prenatally and postnatally to low doses of mercury. *Bull. Environ. Contam. Toxicol.* **3,** 73–79.

100. Prasad, K. N., Nobles, E. and Ramanujam, M. (1979) Differential sensitivities of glioma cells and neuroblastoma cells to methylmercury toxicity in cultures. *Environ. Res.* **19,** 189–201.

101. Rakic, P. (1972) Mode of cell migration to the superficial layers of fetal monkey neocortex. *J. Comp. Neurol.* **145**, 61–84.
102. Rakic, P. (1975) Timing of major ontogenetic events in the visual cortex of the rhesus monkey. In: N. A. Buchwald and M. Brazier (Eds.), *Brain Mechanisms in Mental Retardation*, Academic Press, New York, pp. 3–40.
103. Rakic, P. and Sidman, R. L. (1969) Telencephalic origin of pulvinar neurons in the fetal human brain. *Z. Anat. Entwickl. Gesch.* **129**, 53–82.
104. Rakic, P. and Sidman, R. L. (1968) Supravital DNA synthesis in the developing human and mouse brain. *J. Neuropathol. Exp. Neurol.* **27**, 246–276.
105. Ramel, C. (1969) Methyl mercury as a mitosis disturbing agent. *J. Jpn. Med. Assoc.* **61**, 1072–1077.
106. Ramel, C. and Magnusson, J. (1969) Chromosome segregation in *Drosophila melangastor. Hereditas* **61**, 231–254.
107. Ramon y Cajal, S. (1980) Sur l'origine et les ramifications des fibres nerveuses de la moelle embryonnaire. *Anat. Anz.* **5**, 85–95; 111–119.
108. Rodier, P. M. and Reynold, S. S. (1977) Morphological correlations of behavioral abnormalities in experimental congenital brain damage. *Exp. Neurol.* **57**, 81–93.
109. Rodier, P. M., Reynolds, S. S. and Roberts, W. N. (1979) Behavioral consequences of interference with CNS development in the early fetal period. *Teratology* **19**, 327–336.
110. Rosenthal, E. and Sparber, S. B. (1972) Methylmercury dicyandiamide: retardation of detour learning in chicks hatched from injected eggs. *Science* **11**, 883–892.
111. Schalock, R. L., Brown, W. J., Kark, R. A. P. and Menon, N. K. (1981) Perinatal methylmercury intoxication: behavioral effects in rats. *Dev. Psychobiol.* **14**, 213–219.
112. Schmechel, D. E., Brightman, M. W. and Marangos, P. J. (1980) Neurons switch from non-neuronal enolase to neuron-specific enolase during differentiation. *Brain Res.* **190**, 195–214.
113. Sidman, R. L. and Rakic, P. (1973) Neuronal migration, with special reference to developing human brain: a review. *Brain Res.* **62**, 1–35.
114. Skerfving, S. (1974) Methylmercury exposure, mercury levels in blood and hair and health status of Swedes consuming contaminated fish. *Toxicology* **2**, 3–28.
115. Skerfring, S., Hansson, K. and Lindsten, J. (1970) Chromosome breakage in human exposed to methyl mercury through fish consumption. *Arch. Environ. Health* **21**, 133–139.
116. Snyder, R. D. (1971) Congenital mercury poisoning. *N. Engl. J. Med.* **284**, 1014–1016.
117. Spyker, J. M. and Smithberg, M. (1972) Effects of methylmercury on prenatal development in mice. *Teratology* **5**, 181–190.
118. Spyker, J. M., Sparber, S. B. and Goldberg, A. M. (1972) Subtle consequences of methylmercury exposure: behavioral deviations in offspring of treated mothers. *Science* **177**, 621–623.

119. Su, M. and Okita, G. (1976) Behavioral effects on the progeny of mice treated with methylmercury. *Tox. Appl. Pharmacol.* **38,** 195–205.
120. Su, M. and Okita, G. (1976) Embryocidal and teratogenic effects of methylmercury in mice. *Tox. Appl. Pharmacol.* **38,** 207–216.
121. Takeuchi, T., Morikawa, N., Matsumoto, H. and Shiraishi, A. (1957) Pathologic study of Minamata disease in Japan. *Acta Neuropathol.* **2,** 40–57.
122. Takizawa, Y. (1979) Epidemiology of mercury poisoning. In: J. O. Nriagu, (Ed.), *The Biogeochemistry of Mercury in the Environment,* Elsevier, North-Holland Biomed Press, Amsterdam, pp. 325–365.
123. Tsubaki, T. (1968) Organic mercury poisoning in Agano area. *Clin. Neurol.* **8,** 511–520.
124. Volpe, J. J. and Adams, R. D. (1972) Cerebro–hepato–renal syndrome of Zellweger: An inherited disorder of neuronal migration. *Acta Neuropath.* **20,** 175–198.
125. Wallace, R. A., Fulkerson, M., Shuits, W. D. and Lyons, W. S. (1971) Mercury in the Environment. Oak Ridge National Library, ORNL NSF-Ep-1.
126. Wannag, A. and Skjerasen, J. (1975) Mercury accumulation in placenta and fetal membranes. A study of dental workers and their babies. *Environ. Physiol. Biochem.* **5,** 348–352.
127. Webster, W., Shimada, M. and Langman, J. (1973) Effects of fluorodeoxyuridine on developing neocortex of the mouse. *Am. J. Anat.* **137,** 67–86.
128. Weis, P. and Weis, J. (1977) Methylmercury teratogenesis in the killifish, Fundulus heteroclitus. *Teratology,* **16,** 317–326.
129. Wilson, J. G. (1973) *Environment and Birth Defects.* Academic Press, New York.
130. Wood, J. M., Kennedy, F. S., and Rosen, C. G. (1968) Synthesis of methylmercury compounds by extracts of a methanogenic bacterium. *Nature* **220,** 173–174.
131. Yang, M. G., Krawford, K. S., Garcia, J. D., Wang, J. H. C. and Lei, K. Y. (1972) Deposition of mercury in fetal and maternal brain. *Proc. Soc. Exp. Biol. Med.* **141,** 1004–1007.
132. Yoshino, Y., Mosai, T. and Nakao, K. (1966) Biochemical changes in the brain in rats poisoned with an alkylmercury compound, with special reference to the inhibition of protein synthesis in brain cortex slices. *J. Neurochem.* **13,** 1223–1230.

Chapter 8

Aluminum Neurobehavioral Toxicology

Ted L. Petit

I. Introduction

Humans are naturally exposed to high amounts of aluminum in the natural environment. Aluminum is the third most common element in the earth's crust, and the most abundant metal: it comprises 8% of the earth's crust, with only oxygen (49.5%) and silicon (26%) occurring more commonly. Aluminum is usually found combined with oxygen or silicon, and generally occurs in its oxidized form as alumina, or as a silicate in feldspars and micas. Large quantities of aluminum are found as bauxite in tropical and subtropical areas, but the amounts of this metal biologically available through plants and herbivorous animals is unknown. Aluminum also occurs in varying amounts in fresh water, depending on the local geochemistry, and in large amounts in sea water, and in the air as dust particles containing aluminosilicates.[45,82]

In addition to naturally occurring aluminum, we are exposed to additional aluminum in the form of antacids, cosmetics, beverage containers, drinking water [because of the use of $Al_2(SO_4)_3$ during purification procedures], plants and animals (aluminum is found in plants and animals in proportion to the amount of aluminum available in their environment).

Ted L. Petit: Department of Psychology and Division of Life Sciences, University of Toronto, West Hill, Ontario, Canada

Little information was available on aluminum neurotoxicity prior to the 1960s. Interest in aluminum has increased recently, particularly with the possibility of its involvement in the human diseases, Alzheimer's disease, and dialysis dementia (discussed below).

2. Entry and Metabolism

2.1. Normal Aluminum Levels

Historically, aluminum has not been considered an important toxicant. With the total body burden of aluminum less than 30 mg, it was thought that the skin, gastrointestinal tract, and lungs were effective barriers in excluding environmental aluminum from internal entry.[1,11,84]

In healthy adults aluminum levels are usually under 10 mg/kg dry weight for heart, spleen, liver, muscle, bone, skin, and brain. Aluminum levels in the lungs are generally the highest seen in the body (up to 100 mg/kg) and likely represent an accumulation of airborne aluminum.[1] With aging, aluminum levels do not change significantly with the exception of lung and brain, which show a significant increase with age.[1,59]

Brain aluminum content in the human is around 2 μg/g dry weight, with an upper limit of 4 μg/g considered the cutoff point for normal human brain.[22] The aluminum content of other adult mammalian brains is in the same range.[22]

2.2. Absorption

A number of researchers have now reported that aluminum can be absorbed from the gastrointestinal tract in normal mammals.[45] Berlyne et al.[7] studied aluminum administration via the drinking water in normal, control, and nephrectomized (uremic) rats. They observed significantly higher serum aluminum levels in the normal and uremic rats exposed to aluminum compared to the control rats that did not receive any aluminum. In addition, total body aluminum content was elevated in the aluminum-exposed rats compared to the control group, although individual tissue levels of bone, muscle, liver, and heart were not significantly elevated. More recently, Bowdler et al.[9] studied aluminum ingestion in rats in the form of aluminum chloride or a commercial antacid preparation containing aluminum hydroxide. Rats were intubated with the aluminum solutions daily with doses ranging from 0 to 1650 mg aluminum chloride, or the equivalent amount of aluminum in aluminum hydroxide. There was a significant correlation between the dose of aluminum chloride consumed and serum and brain aluminum levels, and between the dose of aluminum hydroxide consumed and brain, but not serum, aluminum levels. These results with rats clearly indicate that increased aluminum ingestion does lead to increased absorption from the gut, and deposition in the body, including the brain.

Further research with normal human subjects has confirmed the work on rats. Cam et al.[10] examined the effects of oral administration of aluminum hydroxide in five normal subjects and found an uptake of aluminum in the body over the period of the study. Gorsky et al.[36] examined aluminum levels in six subjects hospitalized for other than renal or gastrointestinal problems, and like Cam et al.[10] they found a positive aluminum balance in the patients taking oral aluminum. In further research Kaehny et al.[41] examined the effects of oral administration of aluminum hydroxide, aluminum carbonate, dihydroxyaluminum aminoacetate, and aluminum phosphate. The authors found an increase in both plasma aluminum levels and urinary aluminum excretion with aluminum hydroxide, aluminum carbonate, and dihydroxyaluminum aminoacetate, but with aluminum phosphate no increase in plasma aluminum level was detected. The authors suggest that the aluminum phosphate is virtually insoluble at an acid pH. These results suggest that aluminum may be absorbed mainly in the stomach and proximal duodenum. The results are also consistent with aluminum absorption from the gut, which is dose-dependent, and its entry into general circulation and deposition within certain body tissues.

Like most other metals, aluminum competes for absorption from the gut with other elements. Still and Kelley[79] suggest that fluoride competes with aluminum absorption from the gut, i.e., the more fluoride in the diet, the less aluminum is absorbed, and this reduces the probability of aluminum-associated behavioral toxicological conditions. Little is presently known about the functional aspects of other possible competitors with aluminum.

A major method of entry for aluminum into the normal body other than through gastrointestinal absorption has been suggested by Crapper et al.[22] These authors propose that since the highest amounts of aluminum in the body are generally found in the respiratory epithelium, aluminum might enter the brain directly by axoplasmic transport, particularly along the olfactory nerves. They found that the olfactory bulb contained a considerably higher aluminum content than other parts of the human brain.

2.3. Circulation

Once aluminum has been absorbed from the gut of the normal mammal, it enters the general circulation. In the plasma, there appears to be a protein-bound and a nonprotein-bound fraction of aluminum.[6,32,33,54] King et al.[46] reported that plasma aluminum was associated with albumin, an unknown high molecular weight protein, and possibly a low molecular weight protein or some inorganic species. Lundin et. al.[54] also reported that plasma aluminum (approximately 50%) was bound to a high molecular weight protein (molecular weight greater than 8000). Similar results were found by Elliott et al.,[32] who observed that 60–70% of aluminum is bound to a high molecular weight protein, with 10–20% being bound to albumin and 10–30% being ultrafiltrable. These authors found that the ultrafiltrable component decreased when the plasma aluminum level fell below 200 μg/L, and was not observed in normal subjects. The results of these studies support the contention that aluminum in the plasma exists in both ultrafiltrable and protein-bound forms.

Aluminum does have toxic effects on blood. Siebert and Wells[77] reported that aluminum chloride injections in rabbits resulted in anemia, hemolysis, and leukocytosis. More recently, Elliott et al.[31] have shown that aluminum hydroxide ingestion in humans undergoing dialysis is accompanied by gradually falling hemoglobin concentrations in the months preceding the onset of encephalopathic signs. In addition, porphyria can be induced by increased aluminum levels.[50,75]

2.4. Entry into the Brain

Once in the blood, ample evidence now suggests that aluminum can cross the blood–brain barrier to enter the brain. Liss,[52] using newborn rabbits, found that aluminum chloride ingestion by the mother rabbit resulted in increased aluminum content in the mother's milk and in the brains of the suckling rabbits. It is important to remember, however, that the blood–brain barrier may not have been fully formed in the infant rabbits. DeBoni et al.[27] subjected adult rabbits to repeated subcutaneous injections of aluminum. Atomic absorption spectroscopy showed that brain aluminum levels increased from 1.1 μg/g dry weight (controls) to 2.5–47.9 μg/g in injected animals. Using the aluminum-sensitive Morin stain, they found that neutrophils and monocytes within vessels and capillary endothelial cells exhibited intensely fluorescent nuclei, whereas the nuclei of cells of the choroid plexus were not fluorescent. These results, in conjunction with a low aluminum cerebrospinal fluid content, suggested that aluminum gained access to the brain by the vascular route, crossing the blood–brain barrier rather than the blood–cerebrospinal fluid barrier. Alfrey[1] also found that cerebrospinal fluid aluminum levels were not elevated in dialysis dementia patients despite the clear increase in body aluminum burden in these patients, thus supporting the conclusions of DeBoni et al.[27] Sodium fluorescein injection into aluminum encephalopathic rabbits 24 h prior to sacrifice produced no evidence of an increased permeability of the blood–brain barrier.[47] In addition, the usual signs of damage to the blood–brain barrier, such as brain edema, petechial hemorrhages, and ultrastructural changes in capillary endothelial cells and basement membranes, are absent in encephalopathic rabbits.[27] Therefore, the adult blood–brain barrier appears to be permeable to aluminum salts. Related to this, Bowdler et al.[9] observed that the rate of uptake of aluminum into the brain was dependent on the dose level of oral aluminum administered to rats, but not on the duration of the administration.

Several studies have suggested that blood aluminum levels, similar to blood levels of other metals, can serve as an indicator only of circulating aluminum levels, and not total body store. Bowdler et al.[9] found an inconsistent relationship between serum and brain aluminum concentrations; though there was a correlation in some groups of rats exposed to aluminum chloride, there was no correlation in others. These authors point out that serum aluminum concentration is not necessarily a reliable indicator of brain aluminum concentration. As Alfrey[1] points out "....since plasma is a transport system for aluminum it is influenced by active aluminum loading. So even though plasma aluminum will not assist in determining total body aluminum burden it may identify patients who are receiving excess aluminum from some external source" (p. 51).

The amount of aluminum absorbed from the gastrointestinal tract may be dependent on certain factors. For example, Mayor et al.[56,57] report that increased parathyroid hormone treatment causes increased intestinal aluminum absorption and higher brain aluminum concentrations in rats given oral aluminum. They also observed that parathyroid hormone withdrawal in rats resulted in a rapid decrease of brain aluminum concentration independent of dietary aluminum. Those authors suggest that reduced parathyroid hormone aids in reducing brain aluminum accumulation, and that "both Alzheimers disease and dialysis encephalopathy may be explained in part by parathyroid hormone acting on orally ingested aluminum." Alfrey,[1] however, argues against such a beneficial role for parathyroidectomy, pointing out that brain aluminum content, EEG changes, and some clinical symptoms of dialysis encephalopathy persist following parathyroid surgery.[2,5,12,60] Further, he points out that parathyroid surgery was performed on three patients with dialysis encephalopathy without apparent benefit (all died within 8 months of surgery), and that a number of patients have developed dialysis encephalopathy months to years after parathyroidectomy. Alfrey[1] suggests that parathyroidectomy has no effect on either the accumulation or egress of aluminum from the brain. The controversy between Alfrey and Mayor over the role of parathyroid hormone in aluminum absorption has yet to be resolved.

2.5. Excretion

Aluminum in the plasma is excreted predominantly through the urine. Gorsky et al.[36] reported that the amount of aluminum excreted in the urine is very small in comparison to the amount excreted in the feces of normal humans, suggesting that most of the orally administered aluminum passes through the body unabsorbed. However, any absorbed aluminum is normally excreted largely, if not entirely, through the kidneys.[41,48] Further, Kovalchik et al.[48] have reported that bilary excretion of aluminum remains extremely low even in the absence of renal functioning, suggesting that excretion through the bile does not form a significant alternative excretory pathway.

It is now well-established that an increase in body aluminum occurs with renal failure. This appears to result in part from an inability to excrete the small amount of aluminum absorbed from the gastrointestinal tract. In normal mammals aluminum excretion increases during periods of increased aluminum uptake. As King et al.[45] point out, there are no studies that clearly delineate the mechanism of renal excretion of aluminum. They suggest that in order to appear in the urine, aluminum may exist as part of a sufficiently small molecule to allow it to permeate the glomerular membrane.

As pointed out earlier, aluminum appears to occur in both a protein-bound and unbound form in the plasma. The fact that a portion of the aluminum is bound to plasma protein appears to explain why aluminum is nondialyzable, and is also consistent with the evidence that the component of the plasma protein that binds aluminum is saturable.[40] The role of the plasma protein binding component for aluminum was recently summarized by King et al.[45]: "The fact that aluminum at low concentrations is strongly bound to a saturable plasma component (probably protein) would seem to explain the very low excretion rates seen in people with a low plasma concentration. At high plasma aluminum concentrations, the component of plasma that binds aluminum is saturated. This results in increased urinary excretion of aluminum because more plasma aluminum is in unbound form(s) than can be removed by glomerular filtration. These unbound species of aluminum may also explain the high tissue content of aluminum seen in patients on hemodialysis who have elevated aluminum levels since the excess aluminum is in a form that can be taken up by various tissues" (p. 8).

Thus, the research to date suggests that humans are currently exposed to large quantities of aluminum in their environment, but that the body is capable of minimizing aluminum entry, and excreting the little aluminum that does gain entry. However, at very high levels of aluminum exposure, or in disease processes, circulating levels of aluminum can increase markedly and gain ready access to the brain.

3. Localization

3.1. Localization in the Body

Elevated aluminum levels can be found in several body areas following increased aluminum exposure. For example, raised aluminum levels have been found in the plasma, bone, liver, heart, striated muscle, and brain.[1,2,7,60] Liver and bone are the tissues most frequently involved, and have the highest aluminum content.[1]

3.2. Localization in the Brain

The aluminum that accumulates in the brain does so in a number of brain regions. Crapper et al.[21] found elevated aluminum levels in several brain areas examined, but though their reported values show raised aluminum levels in all areas sampled, it was not homogeneously distributed in the tissue. Atomic absorption analysis showed elevated aluminum levels in five areas throughout the cerebral cortex of aluminum-exposed rabbits, as well as in the midbrain, medulla, and cerebellum.[27] Although there were wide variations in aluminum levels between brain areas, there were no suggestions of a selective accumulation of aluminum by a specific brain area, as there is for other metals such as zinc and lead. Several investigators have, however, found higher concentrations of aluminum in gray matter than in white matter.[2,4,57]

The green florescence of the Morin stain can also be used to localize aluminum. Crapper and his colleagues[21] have used this stain and report the presence of aluminum in neurons and glia in several brain regions. DeBoni et al.[27] found aluminum distribution to be patchy, with cells separated by less than 50 μm showing wide variations in the intensity of florescence. However, aluminum does accumulate in most brain cells, and is not restricted to those with neurofibrillary degeneration (see below).[24] Therefore, it appears that aluminum diffuses widely in the brain, binds to all cells, but induces a morphological change in only certain vulnerable neurons, particularly those with large dendritic trees.[15]

3.3. Intracellular Localization

The intraneuronal accumulation and intracellular binding sites for aluminum have also been examined. DeBoni et al.,[27,28] using the Morin stain to examine aluminum binding in rabbits following aluminum injections, found that aluminum binds to the chromatin of neurons, astroglia, microglia, oligodendroglia, and pericytes throughout the brain within 6 h. They did not observe selective binding to any other intracellular constituents including Nissl bodies, ribosomal aggregates, and regions of neurofibrillary degeneration, or on the microtubules of the spindle apparatus in mitotic plant cells. Aluminum appeared over the entire chromosome, but the intensity of the florescence varied along the chromosome. Cell fractionation studies have been carried out by Crapper and DeBoni[18] to determine which nuclear chromatin fractions bind aluminum in aluminum encephalopathic cats. They reported 370, 540, 300, and 330% increases in the weight of aluminum per gram of DNA for the hetero-, intermediate, and light chromatin fractions, respectively. The aluminum observed bound to nuclear chromatin accounted for 80% of the total brain aluminum (see[20] for a review).

Terry and Pena[80] reported that aluminum was concentrated in the region of the neurofibrillary tangles in aluminum exposed rabbits. Similarly, Klatzo et al.[47] used the solochrome Azurine method to stain for aluminum in aluminum phosphate-injected rabbits. The observed "faint, but unmistakably positive blue staining for aluminum in the area corresponding to the presence of the birefringent neurofibrillary tangles in the affected neurons.

On the other hand, aluminum staining was absent in cytoplasmic clearings themselves which were devoid of the neurofibrillary material. Interestingly, some of the neurons which demonstrated aluminum in the cytoplasm also showed positive blue staining of the nucleolus" (p. 190). Galle et al.[35] used an electron microprobe to examine for aluminum in their sections of cerebral tissue from aluminum-treated rats. They observed aluminum in lysosomes and lipofuscin granules containing crystals. Aluminum was not observed in any other part of the brain.

Thus, there are some inconsistencies in the literature concerning the intracellular localization of aluminum in animals. It would appear that much of the aluminum is localized in the nucleus, but there is some disagreement about its localization in the neurofibrillary tangle. Part of the difference in the results may result from different techniques. The Morin stain is not reliable for the selective identification of aluminum, since it also chelates with boron, beryllium, cadmium, gallium, indium, lead, thorium, uranium, and zirconium.[35] On the other hand, King et al.[45] argue that the electron microprobe may not be sufficiently sensitive to pick up nuclear aluminum. It is also possible that aluminum is distributed differently in different species. Galle et al.[35] utilized the rat, a species that does not develop neurofibrillary tangles or an encephalopathy in response to aluminum. King et al.[44] report that they were unable to study aluminum content in the rat with the Morin stain because of the anomalous properties of the tissue in response to the stain. Differences in the subcellular binding of aluminum might account for the species differences in response to aluminum.

Several investigators have examined intracellular aluminum accumulation in humans. In an early study, Terry and Pena,[80] using electron probe analysis, reported being unable to find any aluminum in scans of autopsied human neurofibrillary tangles. However, more recently Perl and Brody[64] examined hippocampal neurons from elderly demented and nondemented individuals with the use of scanning electron microscopy and X-ray spectrometry. They observed aluminum peaks within the nuclear region of 91% of neurofibrillary tangle-containing neurons from senile dementia patients, and in similar such neurons in nondemented patients. They did not, however, observe aluminum in neurons without tangles from either demented or nondemented patients. Examination of the cytoplasmic rather than the nuclear region gave similar results, except that only 10–30% of the tangle-bearing cells had sufficient cytoplasmic aluminum to be detected. These findings support those of Crapper et al.[19,20] who reported that patients with Alzheimer's disease have higher aluminum concentrations, with most of the aluminum occurring in the nucleus and heterochromatic fractions.

In contrast to the nuclear binding of aluminum in senile dementia,[19] the aluminum in dialysis dementia patients has been reported to be primarily cytoplasmic.[19] This is of particular interest, for in dialysis dementia patients, neurofibrillary tangles have not been observed to accompany increased aluminum concentrations. It is tempting to speculate that the differential cellular responses to aluminum might be caused by different binding sites for aluminum within the cell.

4. Neuroanatomical Effects of Aluminum

4.1. Intracellular Effects

Normal nerve cells contain three types of fibrous organelles, neurotubules (25 nm diameter), neurofilaments (10 nm diameter), and microfilaments (6–7 nm diameter). Research on the effects of aluminum has centered on neurotubules and neurofilaments. In normal neurons, neurotubules and neurofilaments are long, straight, thread-like fibers running through and parallel to the long axis of cellular processes. Neurotubules are prominent cytoplasmic constituents, particularly in dendrites of neurons with large dendritic structures such as hippocampal and neocortical pyramidal cells. Studies into their function have suggested that neurotubules provide a cytoskeletal role that is particularly important in allowing the growth and development of cell processes.[68,92] In addition, numerous studies have shown that neurotubules are responsible for the transport of substances within axonal and dendritic systems (axoplasmic and dendroplasmic transport). Neuroplasmic transport is essential for supplying distal processes with constituents necessary for cell maintenance and enzymes necessary for the synthesis of neurotransmitters.[37,73,74]

Increased brain aluminum accumulation is known to be associated with neurofibrillary tangles in some species. The terms "neurofibrillary tangle" or "neurofibrillary degeneration" refer to an accumulation of tangles of neurofibers within the neuron. Electron microscopic examination of aluminum-induced neurofibrillary tangles indicates that they consist of areas of cytoplasmic clearings, filled with either compact orderly bundles or accumulations of large swirls of 10 nm neurofilaments (see Fig. 1).[17,67,76,80,90] Terry and Pena[80] and Wisniewski et al.[90] report that these 10 nm filaments sometimes appear to have a hollow lu-

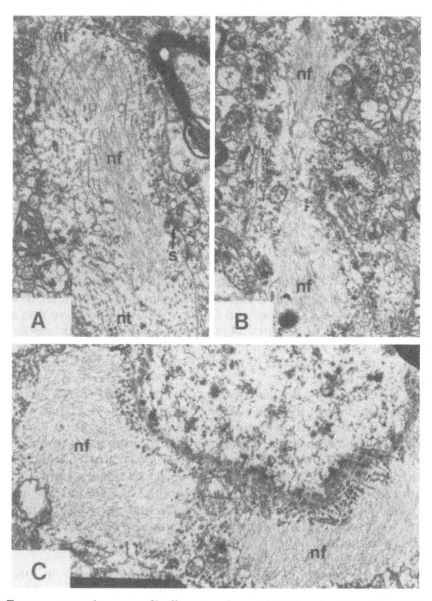

FIG. 1. Note the neurofibrillary tangles within the dendrite (A and B) and soma (C). These tangles can be seen continuous with cytoplasmic areas containing neurotubules. A synapse (S) can be seen on the dendrite in A. 9700×.

men. The lumens, however, were poorly defined and rarely apparent on cross-section. Crapper,[15] however, describes them as solid filaments. The filaments also appear to have fine side arms, 20–50 nm long and about 5 nm wide, extending from the edges of the filaments and interconnecting them. The filaments lie parallel within the bundles that comprise a tangle, and are of indeterminate length.[80]

The tangles are not bound by a membrane, nor do they appear to maintain any topographic relationship with other cell organelles. The tangles are, however, sharply demarcated from the rest of the cytoplasm. They generally push other organelles aside, with only clear homogeneous low electron-dense cell sap found between them, identical to that found elsewhere in the cell. Occasionally other cell organelles, e.g., a few mitochondria, lysosomes, ribosomes, or a small fat droplet, can be found within a tangle.

Organelles outside of the tangle generally appear morphologically normal, although there is some suggestion that they are pushed aside and crowded. Several investigators[17,76,80] have observed intracellular crowding, visible swelling of the soma, and distention of cellular processes by neurofibrillary tangles. Embree et al.[34] reported an increase in organic mass in tangled neurons and Selkoe et al.[76] observed a 20% increase in protein per cell. In a quantitative study, Crapper and Dalton[17] showed that there was a decreased density of normal 24 nm neurotubules in cell processes containing neurofibrillary tangles. In addition, nucleolar diameter and volume are reduced in neurons and glia exposed to aluminum in vitro.[18]

Neurofibrillary tangles are generally found within the soma, frequently close to the nucleus. Selkoe et al.[76] observed tangles on both sides of the nucleus (a not uncommon observation at the light microscopic level) suggesting either a multiplicity or a continuous concentric configuration around the nucleus. Petit et al.[67] and Crapper and Dalton[16] have also observed tangles extending into dendritic processes. The latter authors noted tangles frequently in processes 3–5 μm in diameter, and also noted a high incidence of tangles in dendrites less than 2.5 μm in diameter. Thus, tangles are generally found only in the soma and dendrites, and only in neurons—glia are not affected.[47,80]

Neurochemical analysis of the neurofilaments in the aluminum-induced tangles indicates that they are composed of three proteins with molecular weights of 68,000, 160,000, and 200,000 daltons.[76] The protein bands from the aluminum-induced neurofilaments were significantly augmented at 68,000 and 160,000 daltons compared to the controls. The 200,000 dalton band was not visible at all in the control gels. An antibody raised against the 160,000 dalton component of a modified bovine brain filament preparation produced fluorescent staining of the aluminum-induced neurofibrillary tangle. This component appears to resemble in certain respects normal neurofilament.[3,71]

Not all animals, however, develop neurofibrillary degeneration. The rat, for example, does not develop neurofibrillary tangles,[44,57] even at doses 5–6 times that adequate for the cat[44] (see below for further discussion).

At present there are two major theories that would account for the formation of neurofibrillary tangles. The first suggests that the tangles are somehow caused by a disruption of the process of normal formation or maintenance of neurotubules.[17,83] The findings of diminished numbers of neurotubules in the area of neurofibrillary tangles would support this hypothesis. A second possibility, and one that is not necessarily mutually exclusive of the first, is that aluminum, by binding onto DNA, alters and possibly increases protein synthesis. Such altered protein synthesis might result in the production of the neurofilamentous tangles.[15]

4.2. Light Microscopic Observations

Preparation of tissue for light microscopic observation allows a different picture of the neurofibrillary tangle. Cresyl violet, PAS, and Congo Red staining yields neurons with large multiple, focal pale-blue cytoplasmic "clearings" that are generally grouped around the nucleus.[47,76] Because of their strong affinity for silver, neurofibrillary tangles can be readily observed with the Bodian and Bielschowsky stains. They also lightly stain with the Sudan III stain.[47] In the Bodian and Bielschowsky stains, the normally delicate network of neurofibrils appears as dense tangles.

Light microscopic examination allows a clear understanding of the time-course for the development of neurofibrillary tangles. The exact time course for the development depends on the various brain areas examined in relation to the site of aluminum injection.[88] Wisniewski et al.[88] observed neurofibrillary changes in the spinal cord of aluminum-injected rabbits as soon as 24 h after the injection, with changes in the brainstem and cerebral cortex becoming apparent 3 and 5 d, respectively, after the injection. Crapper[14,15] has outlined the progressive formation of the tangles in rabbits and cats injected with aluminum. He notes the earliest changes in neocortical pyramidal cells 96 h after injection, when thickened and darkened argentophilic neurofibrillary strands are observed in the proximal segments of apical dendrites. In rabbits, after 96 h the strands occupy much of the soma and spread into the apical and basilar dendrites. In cats, although tangles are seen by 96 h and by 6 d the argentophilic material fills the proximal portion of dendrites, tangles are rarely seen in the soma at this time. By 10 d in the cat, strands pass through the cell body and appear to fuse with strands arising in the dendrites. From 6 to 16 d in the rabbit, the strands extend extensively through the soma, the proximal portions of dendrites, beyond the bifurcation of the apical dendrite, and occasionally into smaller branches.

Different regions of the brain and different cell types appear to be differentially susceptible to the development of neurofibrillary tangles. This differential sensitivity was extensively discussed by Klatzo et al.,[47] Wisniewski et al.,[88,89] and Crapper and Dalton.[17] In the spinal cord, the anterior horn cells throughout the whole length of the cord show the most pronounced neurofibrillary degeneration, with the cervical and lumbar regions being most affected with intracerebral or cisternal injections.[47] Selkoe et al.[76] reported that the greatest density of affected cells occurs around the central canal and in the adjacent gray matter within 1 mm on either side, where 75% of neurons (largely of intermediate size) are affected. The proportion of affected cells declined toward the depth of the anterior horns, with the largest anterior horn cells (presumably alpha moroneurons) infrequently affected. However, Wisniewski et al.[90] reported anterior horn cells were particularly susceptible with dorsal horn neurons showing little degeneration. Wisniewski et al.[88,89] similarly noted that cells of the substantia grisea intermedia and nucleus motoris medialis were the most affected, whereas the nucleus motoris lateralis showed little change. In the brainstem, Selkoe et al.[76] observed that the gray matter of the brainstem tegmentum was uninvolved, but neurofibrillary tangles were seen in the nuclei of the basis pontis. Klatzo et al.[47] and Wisniewski et al.[88,89] noted affected cells in the olivary system, nuclei raphe, nucleus magnocellularis, nucleus papilliformis, spinal trigeminal nucleus, dorsal motor nucleus of the vagus nerve, nucleus cochlearis, corpus quadrigeminum anticum and posticum, substantia grisea, red nucleus, and formatio-reticularis. However, neurons in the hypoglossal, facial, and trigeminal nuclei, the nucleus cochlearis ventralis, red nucleus, inferior colliculus, superior colliculus, and reticularis nucleis showed tangle involvement that was inconsistent and varied between animals.[47,88,89] The nucleus prepositos hypoglossi, subnucleus Deiters, and the mesencephalic nucleus of the fifth nerve appeared resistant to tangle formation.

Additional neurofibrillary tangles were seen in the ventral nucleus of the thalamus, the lateral, and less frequently, the medial geniculate nucleus, and mamillary bodies. The zona incerta and substantia nigra were only mildly affected, whereas the caudate, globus pallidus, putamen, claustrum, and medial nucleus of the thalamus were free of tangles. Purkinje cells of the cerebellum were regularly affected. In the hippocampus, the pyramidal cells of the CA areas are extensively involved as are pyramidal cells of the entorhinal area.[17] However, dentate granule cells remain unaffected. In the cerebral cortex, the pyramidal cells of layers III and V are most frequently affected, followed by the pyramidal cells of layers II and VI, with layer IV granule cells showing no changes.[14,15] Some researchers have found changes in the cerebral cortex difficult to produce[47,76,90] or report that tangle formation within cortical areas is usually patchy.[17] Although widespread areas of the central nervous system are affected, certain cells appear to be more or less resistant to the formation of neurofibrillary tangles. The cells with large dendritic structures, large to moderate cell bodies, and large diameter proximal dendrites are more prone to tangle formation.[20,89] In addition, the tangle distribution may reflect injection site and individual cerebral spinal fluid flow.[17,89] Aluminum can induce in vitro changes in human cortical cells, but not in cells of the dorsal root ganglion.[29] Interestingly, despite high aluminum concentration in human dialysis encephalopathy and Alzheimers disease, in vivo neurofibrillary tangles identical to those produced by aluminum in other species have never been reported in humans.

There does appear to be some correlation between the amount of aluminum in an area and the presence or absence of neurofibrillary tangles. In the cat brain, areas with aluminum concentrations below 4 µg/g dry weight do not show neurofibrillary degeneration. In areas with aluminum concentration greater than 4–6 µg/g dry weight, the density of cells with tangles is approximately proportional to the aluminum concentration.[24] The concentration of aluminum found to induce an encephalopathy in rabbits is in the range of 6–12 µg/g following intracranial injections.[27] In the rat, however, concentrations of up to 40 µg/g do not induce neuropathological effects (see below for discussion on species differences).

Aluminum appears to have deleterious effects on the dendritic structure of the cell. Petit et al.[67] used the Scholl analysis to

examine the number of dendritic branches at 20 μm intervals from
the cell body in layer V pyramidal cells from the sensorimotor cortex of aluminum-exposed rabbits. They observed a sharp and progressive decrease in the number of dendritic branches with
increasing distance from the cell body in the aluminum-exposed
animals. This pattern is consistent with that expected from a dying
back process. Westrum et al.[87] observed a loss of dendritic spines,
dendritic beading, and degeneration after topical alumina cream
application. Thus, aluminum-induced neuronal changes may not
be compatible with long-term dendritic health. We have recently
suggested that reduced neuroplasmic transport may underlie this
cell atrophy. The accumulation in aluminum-treated neurons of
large bundles of haphazardly oriented neurofilaments pushing
other cell organelles aside, along with a concomitant reduction of
neurotubules suggests that neuroplasmic transport would be
disrupted in that area of the cell. The disruption of normal neurotubule structure or function by other agents results
in the blockage of the fast component of neuroplasmic transport
in axonal and dendritic systems.[25,73,78] A disruption of neurotubules or neuroplasmic transport results in a diminution of
transmitter substances, or enzymes necessary for their synthesis,
as well as of cell constituents necessary for cell maintenance. We
suggested that aluminum-induced neurofibrillary tangles might
reduce neuroplasmic transport resulting, in part, in the altered
electrophysiological properties and dying back of distal cell processes observed in aluminum-treated animals. In an important
study, Liwnicz et al.[53] examined anterograde and retrograde
axonal transport in aluminum-treated rabbits. They injected radioactive leucine into the area of the spinal motoneurons (anterograde transport) or the tongue (retrograde transport) and measured its transport over time. Although retrograde transport was
blocked in some, but not all animals, they found no effect of aluminum on anterograde axoplasmic transport. These results suggest
that although aluminum may affect retrograde transport, it does
not disrupt anterograde transport in axons of neurons with cell
bodies in neural areas containing tangles. However, as the authors
point out, the transport waveform used for calculation represents
a composite of the most rapidly transported material in most fibers in thousands of axonal profiles. Unaffected neurons can contribute substantially to the profile, and "the point would still exist
if the majority of the fibers had completely ceased their transport
activity." However, perhaps more important for understanding
dendritic atrophy, neurofibrillary tangles are typically seen only in
the soma and dendrites, whereas Liwnicz et al.[53] measured trans-

port in axons where tangles are not seen. From our knowledge of the intraneuronal location of the tangles, disruption in transport might reasonably be expected in dendritic systems of cells with large dendritic trees where the greatest tangle formation is generally seen. Measurement of dendritic transport in such neurons would greatly aid in understanding the mechanism of aluminum action.

Few other light microscopic changes have been noted. In particular, researchers have observed no abnormalities in blood vessels, glial elements, myelin, or elements of the blood–brain barrier.[27,47] Neurons appear to be selectively effected.

4.3. Gross Anatomical Observations

Alterations in the gross structure of the brain accompanying intracranial injection of aluminum are minimal if existent at all. Klatzo et al.[47] suggested a possible decrease in brain volume, based on a decrease in water content and chloride space, but this has not been documented. Wisniewski et al.[88] observed hydrocephalus in some of their animals; however, they did not report neutralizing the pH of their solution. We (unpublished observations) have observed similar hydrocephalus when nonneutral solutions of aluminum are injected; this is not observed with neutral solutions. Presumably the highly acidic solution damages the normal cerebrospinal fluid resorptive tissues.

5. Electrophysiological Effects of Aluminum

5.1. Membrane Effects

Blaustein and Goldman,[8] using the voltage clamp technique, studied the effects of aluminum on action potential generation in lobster axons in vitro. When aluminum in concentrations as low as 0.03 mEq/L was substituted for calcium in the bathing medium, the action potential was blocked. They also observed a significant effect when aluminum was added to normal seawater. Aluminum caused a marked reduction of early transient current and also reduced the steady-state current.

5.2. Single Unit Responses

Crapper and Tomko[24] examined the single unit response of visual neocortical neurons in aluminum-treated cats. Spike discharges were collected for 10–120 min and the mean spontaneous frequency of discharge was calculated. There was a significant decrease in the number of neurons with spontaneous frequencies between 7 and 10 spikes/s and an absence of neurons with spontaneous mean frequencies above 14/s in the aluminum-treated animals. The authors argue that the recordings were probably from pyramidal cells and suggest that the effect of aluminum is to render them electrically inactive.

They also observed a 10% reduction in the percentage of neurons with spontaneous mean frequencies less than 2.5. They interpret their finding, in light of other evidence, as suggesting that aluminum, at least in its early stages, reduces the spontaneous discharge rate.

Neuronal discharge analysis during visual stimulation revealed that in neurons of aluminum-treated cats there is a marked increase in the average mean frequency of evoked spike discharge.[24] In addition, the neurons exhibited significantly larger changes in the average hit probability with stimulation, and a change in the probability of discharge with stimulation. Further, there was a reduction in the number of neurons with a short latency response to the stimulation, and a smaller percentage of neurons with nonstationary spike discharges compared to a simple Poisson process.

A number of other neuron or membrane electrophysiological characteristics have been examined. The spike properties of spinal motor neurons in aluminum-treated rabbits examined during intracellular recording are not different from controls in terms of antidromic conduction velocity, spike rising and falling phase, spike duration, and action potential amplitudes.[15] Intracellular recordings from the cerebral cortex of aluminum-exposed rabbits during the early and middle stages of the encephalopathy are similar to controls in terms of excitatory and inhibitory postsynaptic potential (IPSPs and EPSPs) amplitudes, discharge threshold, action potential rheobase and efficiency of cortical inhibition. However, in the late stages of the encephalopathy, the majority of the cells exhibit alterations in the normal firing patterns evoked by transcallosal stimulation, and show prolonged episodic depolarizations.[14] Peripheral nerve conduction velocity in the sciatic nerve was also measured in the aluminum-exposed rabbit and found to be normal, as were electromyogram recordings.[15] No fasciculations, fibrillations, or unusual action potential waveforms were detected. This led Crapper[15] to conclude that "there was no evidence that soma-dendritic neurofibrillary degeneration affected axon conduction, spike mechanisms, synaptic transmission or induced changes in the postsynaptic muscle membrane."

5.3. Electroencephalographic (EEG) Changes

Studies in both cats and rabbits reveal that aluminum has minimal effects on the EEG recording until late in the encephalopathy, well after behavioral, evoked potential, and single unit electrophysiological changes have been noted.[14,15,17,24] Seven days following aluminum exposure in rabbits, occasional brief bursts of 2–4 cycles/s slow waves occurred in a predominantly normal record. The authors found minimal changes in the EEG until the late stages of the encephalopathy. In the cat, the EEG shows similar slow wave activity of 3–6 Hz appearing infrequently between 12 and 14 days post-injection. The EEG became disorganized with large amplitude 1–3 Hz slow waves around day 24 of the encephalopathy. Similar recordings from the hippocampus revealed no alterations in electrical activity during the 15 d post-injection recordings.

Further examinations have been made on the sleep–wakefulness cycle of aluminum-exposed cats. The amount of time spent in slow-wave sleep went from 58% prior to injection, to 63% on day 12, and 72% on day 15 post-injection. Conversely, rapid eye movement (REM) sleep went from 13% pre-injection to 14% on day 12, and only 7% on day 15 post-injection.

5.4. Evoked Potentials

Visual evoked potentials (VEP) have been studied following aluminum infusion in both cats and rabbits. In rabbits the progressive effect of the aluminum encephalopathy was investigated by serial comparisons of averaged evoked potentials in the same animals with chronically implanted electrodes.[14,15] The evoked potential began to decline slowly between days 5 and 7 and then rapidly declined 3 d prior to death. The short latency initial positive wave, of 15 ms latency, showed no alteration through the course of the encephalopathy. The later components of the visual evoked potential, i.e., the large negative wave of peak latency of about 130 ms, and the positive wave of peak latency of about 300 ms, declined after 7 d and virtually disappeared 48–96 h prior to the onset of status epilepticus. Animals in which aluminum was slowly injected directly into only one hemisphere showed similar visual evoked potential declines in the injected, but not the noninjected side. Both the resistance of the initial short latency wave and the further finding that the optic tract response to optic nerve stimulation was normal suggest that the VEP changes are a result of neocortical changes.

In rabbits, similar changes in the VEP have been observed.[17] Few or no changes were observed during the first 14 d post-injection. With the onset of neurological signs, there was a trend toward longer latency in the early, P1, N1, and P2 peaks of the lateral gyrus and in all peaks from the suprasylvian gyrus.

Transcallosal stimulated evoked responses have also been examined in the cat.[14,15] There is a progressive decline in the surface negative wave after day 7 post-injection that then reversed polarity and was replaced by a large surface positive wave 24–72 h prior to the onset of status epilepticus. The late, 120 ms peak latency, surface positive wave also reversed polarity in the later stages of encephalopathy.

6. Neurochemical Effects of Aluminum

6.1. Transmitter Systems

Little systematic evidence exists concerning the effects of aluminum on neurotransmitter systems. The cholinergic system has received the most attention. Miller and Levine[62] found that acetylcholinesterase was significantly depressed in aluminum-treated mouse neuroblastoma cells in vitro. After 30 min incubation, acetylcholinesterase levels fell to 64% of that in control cultures. An aluminum-induced inhibition of acetylcholinesterase has also been reported by others.[49,63] Hetnarski et al.,[39] however, reported no changes in the cholinergic enzymes choline acetyl transferase and acetylcholinesterase in the spinal cord of rabbits with aluminum chloride-induced neurofibrillary degeneration. Aluminum has also been reported to inhibit choline uptake by synaptosomes.[51] Thus, several lines of evidence suggest that aluminum has a disruptive effect on cholinergic functioning.

Little has been done to examine other transmitter systems although aluminum has been shown to have some inhibitory effects on the synaptosomal uptake of dopamine, norepinephrine, and 5-hydroxytryptamine.[51] Wenk and Stemmer[86] investigated the effect, in rats, of aluminum in combination with a number of special diets, many of which were deficient in other metals. The various interactions between the metals were complex and difficult to interpret in terms of the norepinephrine and dopamine levels examined. Animals fed an aluminum-contaminated, "optimal nutrient" diet, showed elevated norepinephrine levels in the cerebral cortex and hippocampus, while animals fed aluminum-contaminated laboratory chow showed reduced norepinephrine levels in the cerebellum. In addition, dopamine levels were increased in the hippocampus in rats fed added aluminum.

Thus, the effects of aluminum on transmitter systems is only just beginning to be explored. As many of the electrophysiological effects appear to be correlated with the degree of neurofibrillary degeneration, it would be of interest to know if the same were true of neurotransmitter changes. This is particularly important when using animals, such as rats, that do not develop an encephalopathy or neurofibrillary tangles and comparing them with animals such as the cat and rabbit that do develop these changes.

6.2. Other Neurochemical Changes

Aluminum has been reported to inhibit enzyme systems involving ATP and to inhibit Na-K-ATPase.[38,51,82,91] Further, aluminum appears to inhibit hexokinase, apparently by complexing with ATP.[38,82,91]

Additional research has centered on the effects of the previously discussed nuclear binding of aluminum. Aluminum is bound in neurons to the acid protein–DNA complex. Aluminum can alter the reassociation of DNA strands following thermal denaturation.[42,43] Depending on the pH, two DNA–aluminum complexes appear to exist that show different effects. At the higher pH range (> 6) aluminum stabilizes DNA to thermal denaturation as a result of binding to the DNA phosphate, while at lower pH levels, aluminum results in base binding and destabilizes DNA, as evidenced by a reduction in the temperature at which the double helix is unwound. Further, aluminum induces changes in the number of sister chromatid exchanges in lymphocytes in vitro, and increases the incorporation of thymidine in astrocytes in vitro.[29] The latter result may reflect DNA repair secondary to DNA damage by aluminum. Miller and Levine[62] reported that RNA content was 20% less than control in neuroblastoma tissue cultures exposed to aluminum. Thus, among other things, aluminum may be interfering with the transcription of DNA, and bringing about altered protein synthesis.

7. Behavioral Effects of Aluminum

7.1. Pre-encephalopathic Changes

The behavioral effects of aluminum have not been extensively studied. Crapper and Dalton[17] described a number of behavioral changes in cats after infusion of aluminum chloride into the brain.

They observed subtle changes in motor coordination 7 d post-injection. Animals were trained to jump from a height of 75 cm onto a platform that measured the sequence of limb impacts. At approximately 7 d post-injection their impact posture began to change, with the cats failing to break the fall by forelimb flexion. The difficulty in motor coordination increased such that by 12–18 d most animals landed so awkwardly that they refused to jump spontaneously.

A progressive impairment in performance on a delayed response task was also seen in these cats. The animals were required to choose the previously observed baited cup in a modified Wisconsin general test apparatus following 10 s delay in which an opaque screen was lowered blocking the view of the cups. Deficits were first observed at 10–12 d post-injection on the longest delay intervals of 30–40 s. Over the next 4–6 d performance fell to chance levels with progressively shorter delay times, although the animals could still perform the task with no delay. These results suggest that aluminum can interfere with short-term memory when visual and motor functions necessary to perform the task are still intact. Cats were also tested 9–10 d after aluminum infusion on a conditioned (one-way) active avoidance task in which the cats had to learn to escape from a chamber to avoid shock. The aluminum-treated animals required more trials to reach criterion on this task than the control animals. In addition, four cats were implanted with electrodes in their lateral hypothalamus, trained to self-stimulate, and then placed on a fixed-interval schedule of reinforcement. The pattern of lever pressing did not show any substantial alterations until late in the encephalopathy, with the onset of grand mal seizures. Thus, several behavioral functions including learning and memory functions undergo a decline early in the encephalopathy well before the onset of overt neurological signs. The same authors reported a correlation between the number of trials to reach criterion on the conditioned avoidance task and the amount of neurofibrillary degeneration in the neocortex and entorhinal cortex.[17] EEG recordings taken on three cats during the delayed response task indicated that slow-wave disorganization was not seen until several days after performance had reached chance levels.[15] Therefore, the behavioral changes were noted after forebrain neurofibrillary degeneration was present, but before gross electrophysiological disorganization.

Similar behavioral results were more recently reported by Petit et al.[67] in rabbits. Ten days after aluminum infusion, rabbits were trained on a step down active avoidance task, i.e., they were taught to step down from an elevated platform to avoid an electric shock. They were tested for retention of the task 13 d following infusion. The aluminum-treated animals were deficient on original learning as well as on retention of the task, requiring more trials than controls to master both. At sacrifice these animals were found to have the neurofibrillary degeneration and dendritic atrophy previously discussed.

Additional behavioral research has centered on rats, a species that does not develop neurofibrillary degeneration. King et al.[44] tested aluminum chloride-infused rats on a conditioned (one-way) active avoidance response similar to that employed by Crapper and Dalton[17] in cats. They did not observe any consistent effect of aluminum on the acquisition of this task. Further, the authors report that even at brain concentrations 5–6 times that found in experimental cats the aluminum injected rats did not show evidence of neurofibrillary degeneration, learning and memory loss or a progressive encephalopathy. Bowdler et al.,[9] however, observed behavioral alterations in rats fed aluminum chloride. They examined open-field activity level, one-way active avoidance acquisition, and sensor-motor abilities on the roto-rod, and recorded the electro-retinogram during visual stimulation. They observed no significant differences between groups on the majority of tests. They did see, however, a reduction in the length of time the aluminum rats could maintain roto-rod activity, an increase in the total distance traveled by the aluminum rats in the open field, and an increased sensitivity to flicker. It is difficult to reconcile these findings with those by King et al.[44] although the differences might be caused by different aluminum dose levels, different behaviors measured, or different modes of aluminum administration. It is unfortunate that Bowdler et al.[9] did not cite the King et al.[44] article and discuss the discrepant findings.

Bowdler et al.[9] also examined the effects of increased aluminum intake in otherwise normal (nondialysed) humans. They examined subjects on the Trail Making Test, the Serial Sevens Test, the Digit Symbol, Block Design, and Digit Span Subtests of the Wechsler Adult Intelligence Scale, and critical flicker frequency. The subjects with the higher aluminum intake showed impaired visuo-motor coordination, poor long-term memory, and increased sensitivity to flicker.

7.2. Encephalopathic Changes

Although subtle behavioral changes are noted in animals soon after aluminum infusion, with increasing time the behavioral changes become more severe as overt neuropathological signs appear. In the rabbit the symptoms generally appear first between 8 and 16 d post-infusion and have been described by several authors.[14,47,76,80] Most authors describe a progression of symptoms; the first to appear is reduced motor activity and an increased irritability to external stimuli. The animals then begin to show a progressive head and body tremor, with a slow gait and the body held close to the ground. An unsteady gait with spread limbs develops with progressive weakness and incoordination beginning in the hind limbs. The animals lose their ability to hop and the righting response becomes weak and eventually disappears. They become ataxic and apathetic, seldom move spontaneously, and some display increased muscle tone and abnormal tonic posturing of the limbs if placed on their side. In the terminal stages, which usually appear between 10 and 28 d post-infusion, animals stop eating and develop focal seizures followed by generalized convulsions. If the animals are not sacrificed, death from starvation or continuous convulsions generally occurs within a few days.

In cats, the general behavioral picture is the same.[15,16] The animals appear normal for the first 8–21 d, with the first sign being a difficulty in maintaining balance and jumping. Crapper and Dalton[16] summarized the final stages: "Over the next 7–14 days a progressive deterioration in movement associated initially with a motor dyspraxia and followed by a truncal ataxia, head tremor, difficulty in maintaining balance, apparent loss of curiosity and transient diarrhea with movement induced vomiting occurred. This stage was followed by retarded motor activity, apparent apathy, myoclonic jerks and general motor seizures." The authors report that they could maintain some animals with intensive nursing, but a motor defect resembling a sensory-cerebellar ataxia remained with a sensory-evoked decorticate rage-like reaction.

As pointed out earlier, some species such as the rat, fail to develop an encephalopathy. Even when observed for 12 months after injection of massive amounts of aluminum, no motor signs of an encaphalopathy appear.[44]

8. Implications in Human Disease

Increased aluminum levels have been implicated in several human diseases. Two diseases, however, Alzheimer's disease and dialysis encephalopathy or dementia, have received the greatest attention.

8.1. Alzheimer's Disease

Until recently it was commonly assumed that senile dementia was caused by arteriosclerosis. We now realize, however, that the majority (50–70%) of cases of senile dementia are caused by Alzheimer's disease, with another 15–20% of the cases of senile dementia having both vascular and Alzheimer's deterioration. The impression of most investigators is that the Alzheimer's changes are responsible for the behavioral dementia (see Petit[66] for a review). Alzheimer's disease has an increased frequency of onset beyond the age of 65 and has a very short life expectancy, with 95% of the patients dead within 5 years; it is the fourth largest killer in North America.

Behaviorally, Alzheimer's disease is characterized by a progressive loss of higher mental functioning.[66] The patients have learning and memory difficulties, show personality changes, and become slowly more debilitated with poor personal hygiene and language difficulties. As the disease progresses patients require total nursing care, losing urinary and fecal control, and showing a profound loss of cognitive abilities. Death usually results from secondary infection. Neuropathological changes seen in Alzheimer's disease include, among others, an accumulation of lipofuscin, neurofibrillary tangles, granulovascular changes with dendritic atrophy, senile plaques, and cell loss (see Petit[66] for review). The presence of neurofibrillary tangles and their similarity to aluminum-induced tangles, along with the behavioral deterioration seen in both syndromes, promoted interest in the possibility of aluminum involvement in Alzheimer's disease. However, the tangles found in aluminum-exposed animals are not the same as those found in Alzheimer's disease. The tangles commonly seen in Alzheimer's disease are abnormal intraneuronal accumulations of paired helical filaments, i.e., paired 10 nm filaments twisted in a double helix with an average periodicity of 80 nm. Double helix twisting is not seen in the aluminum tangles. Nonetheless, a number of other factors have contributed to increased interest in the possible involvement of aluminum in Alzheimer's disease.

The aluminum content in normal brain has been reported to range from 0.1 to 3.9 μg/g dry weight.[22] Mean aluminum content ranges from 0.7 μg/g in fetuses and neonates to 1.8 μg/g in adults. Similar results have been reported by others, e.g., 1.9 μg/g,[22] 1.25 μg/g,[81] 2.2 μg/g in gray matter and 2.0 μg/g in white matter,[2] 1.53 μg/g,[58] and 3 μg/g. From these findings, an average aluminum content above 4 μg/g dry weight appears to be abnormal, i.e., exceeds three standard deviations of the mean.[22]

Increased levels of aluminum have been reported in the aged or Alzheimer brain. The aluminum content of 585 areas in 10 postmortem cases of Alzheimer's disease has been examined, and a range of 0.4–107.0 µg/g was found.[22] Twenty-eight percent of all regions sampled had concentrations greater than 4 µg/g, the level considered the normal upper limit. Further, there was a correlation between increased aluminum content and areas of neurofibrillary degeneration (although not senile plaques) in two Alzheimer's brains. In two vascular dementias, normal aluminum concentrations were found. Biopsy material from two Alzheimer's patients also revealed similar increased aluminum content of 11.5 and 9.9 µg/g dry weight.[21]

The aluminum in the Alzheimer brain appears not to be uniformly distributed,[20-22] i.e., there are large regional differences in the content of aluminum in Alzheimer brains and this may have led to interlaboratory differences in reported results for aged control and Alzheimer's brains. Though Crapper has reported increased aluminum only in Alzheimer's brains, and not in aged controls, McDermott et al.[58,59] found elevated aluminum concentrations in the frontal cortex and hippocampus of elderly normal individuals compared to younger individuals. They also found no difference in aluminum concentration between 10 Alzheimer patients and normal aged controls. Their results suggested that aluminum accumulation might be a normal consequence of increasing age, and not specific to Alzheimer's disease. These differing results have been addressed in a joint study between the two laboratories.[23] Though there was good agreement between the laboratories on brain aluminum content in a multi-infarct (non-Alzheimer's) dementia (2.42 and 2.43 µg/g dry weight), there were marked differences between the laboratories on brain aluminum content of a case of dementia with neurofibrillary tangles (6.02 vs 2.92 µg/g dry weight). The authors suggest the differences are the result of selection of "control" brains, and selection of sample areas, but at present the issue has not been completely settled. Complete resolution of the question will be difficult because of the non-uniformity of aluminum distribution, and the difficulty in selecting "normal" aged controls. Moreover, neurofibrillary tangles are common even in "normal" nondemented aged brain, and in fact, some degree of neurofibrillary tangles are present in virtually all individuals over the age of 80 (see[66]).

Perl and Brody[64,65] examined neurons with X-ray spectrometry from patients with senile dementia. They observed that 91.2% of the neurons with neurofibrillary tangles demonstrated

detectable foci of aluminum within the nucleus, whereas aluminum was detected in only 3.8% of adjacent neurons without
neurofibrillary tangles. Aluminum peaks were also detected within
the nuclear region of 88.9% of neurofibrillary-containing neurons
from elderly nondemented patients, but in only 5.9% of nontangle-
containing neurons.

The results of controlled exposure of human neurons to aluminum in vitro have been reported by DeBoni et al.[29] The dorsal
root ganglion neurons remained uneffected by aluminum exposure, while neocortical cells showed an accumulation of
neurofibrillary tangles with aluminum exposure in the concentration range found in Alzheimer's disease. However, the observed
tangles were an accumulation of 10 nm neurofilaments identical
to those seen with aluminum exposure in animals and not like the
paired helical filaments seen in Alzheimer's disease. Paired helical
filaments have been induced (*in vitro*) in human neurons by the
same researchers, but as a result of exposure to Alzheimer brain
extract, and in the absence of aluminum.[26]

8.1.1. Aluminum and Alzheimer's Disease: Conclusions

For a number of reasons outlined above, it is now fairly clear that
aluminum is not the primary factor in the pathogenesis of
Alzheimer's disease. Although increased aluminum levels are
found in Alzheimer's disease and in particular in tangle-bearing
neurons, a number of findings argue against a primary role: (1) increased aluminum may be a consequence only of advancing years,
(2) elevated aluminum content in human aluminum encephalopathy[61] and dialysis dementia is not associated with Alzheimer neuropathology, (3) aluminum-exposed human neurons in vitro develop single strands of 10 nm filaments, not the paired helical
filaments of Alzheimer's disease, (4) spinal cord and cerebellar
Purkinje cells in aluminum-treated animals show neurofibrillary
degeneration, while tangles are not seen in these cell types in
Alzheimer patients. Other lines of evidence concerning the
neurochemical makeup of Alzheimer's and aluminum tangles, as
well as the cholinergic involvement in each of the syndromes, are
less clear at this time. The reader is referred to arguments by
Wisniewski et al.[88] and Crapper McLachlan and DeBoni[20] for further discussion on the possibility of aluminum playing a role in
Alzheimer's disease. Wisniewski et al.[88] conclude that Alzheimer's
disease "is not caused by aluminum intoxication," while Crapper
McLachlan and DeBoni[20] hypothesize that "an unknown primary
pathogenic event in Alzheimer's disease alters brain aluminum

metabolism so that the metal gains access to brain tissue compartments from which it is normally excluded. Hence, this hypothesis
considers aluminum to be a secondary compounding factor in the
pathogenesis of this untreatable condition." However, they also
are forced to conclude that "there is insufficient evidence to judge
whether aluminum exerts a cytotoxic role in processes associated
with neurofibrillary degeneration of the Alzheimer type, or
whether the metal merely accumulates in a non-reactive form as a
secondary consequence of the disease."

8.2. Dialysis Dementia

Dialysis dementia is a neurological syndrome observed in some
patients undergoing renal dialysis which begins with symptoms of
speech difficulties. It is characterized psychologically by symptoms that include paranoia, personality changes, sometimes severe psychoses, confusion, impaired mathematical abilities, and
delirium. With progression of the disease, a loss of muscle
coordination, marked motor abnormalities, and the onset of seizures occur, the clinical course terminating in convulsions and
death.[2,13,70]

Several factors have implicated the role of increased aluminum in this disease. Alfrey et al.[2] reported elevated aluminum levels in muscle, bone, and brain gray matter in dialysis patients, suggesting that aluminum was the etiological agent in dialysis
dementia. McDermott and Smith[60] correlated the duration of dialysis treatment, using softened or untreated tap water to make up
the dialyzate, with the brain aluminum level in dialysis dementia
patients. Several investigators have shown that dialysis dementia is
associated with high aluminum content in the water used to make
up the dialyzate.[30,70,72] Further, elimination or reduction of the
dementia is observed when deionized water is substituted. Several
authors also observed a reversal of dialysis dementia when oral
aluminum hydroxide intake was stopped.[55,69,72]

Several investigators have reported elevated brain aluminum
content in dialysis dementia since Alfrey et al.[2] first reported such
results.[1,4,12,60] Although brain aluminum content for non-
encephalopathic dialysis patients are reported by different authors to be 3.8–8.5 mg/kg dry weight, brain aluminum content in
dialysis dementia has been reported in the range of 12.4–33.0
mg/kg. Although the aluminum levels in nondemented dialysis patients may be elevated, neurological signs appear when the tissue
content of aluminum reaches or exceeds 10–20 times the content

found in normal brains. Interestingly, the aluminum in neurons of dialysis patients appears to be located in the cytoplasm rather than associated with the nucleus, as is seen in Alzheimer's disease.[19]

Although the aluminum concentration in dialysis encephalopathy may be 2–3 times greater than that observed in Alzheimer's disease,[19] Alfrey et al.[2] found no specific neuropathological changes. In particular despite the high concentrations of aluminum, no evidence of neurofibrillary degeneration was observed. Thus, none of the neuropathological signs of Alzheimer's disease are observed in dialysis dementia as one might anticipate if Alzheimer's disease were the result of elevated aluminum levels.

Therefore, there appears to be considerable evidence to suggest that elevated aluminum may be the precipitating neuropathological event in dialysis dementia, while aluminum may be a secondary event in Alzheimer's disease. That neurofibrillary tangles are not found with the extremely elevated aluminum levels seen in dialysis dementia would appear to be inconsistent with the reported production of such tangles with elevated aluminum in human neurons in vitro. Unfortunately, it is not possible to reconcile these two findings at this time. Clearly, much more work needs to be done to clarify further the possible role of aluminum in human disease.

9. Methodological Considerations

9.1. Anatomical Techniques

The most frequently used method to assay for the neuropathological effects of aluminum is the neuroanatomical verification of neurofibrillary tangles. This can be done either at the light or electron microscopic level.

At the light microscopic level, tissue can be thinly sectioned and examined either with stains for cell constituents other than the neurofibrillary tangles (e.g., cell stains), or with ones that stain the tangles themselves (e.g., silver stains). Affected neurons stained with cresyl violet or toluidine blue show clear areas in the cytoplasm, indicating areas of neurofibrillary tangles. Conversely, staining with the Bielschowski or Bodian stains reveal darkly stained strands of neurofibrillary tangles in the cytoplasm.

At the electron microscopic level, neurofibrillary tangles are easily observed in the soma of neurons fixed in glutaraldehyde and processed through osmium tetroxide.

9.2. Methods and Dosages of Aluminum Administration

The majority of the research on experimental animals has involved intracranial injections of aluminum solutions. In rabbits an encephalopathy has been produced by injectiong solutions of aluminum chloride, aluminum phosphate, aluminum tartrate, or aluminum lactate.[14,47,67,76] Five micromoles of aluminum appears to be within the range suitable for producing a progressive encephalopathy. The injections are generally made by slow infusion of a 100–200 μL volume. The solution should be sterilized by autoclaving to prevent viral contamination, since virus particles are absorbed on salts of aluminum.[85] In addition, the pH should be adjusted to the neutral range to prevent damage to arachnoid villi with resulting hydrocephalus (Petit, unpublished observations).

The same solutions have been used in cats generally in the range of 5–15 μmol. Five to eight μmol appears to be sufficient for production of the encephalopathy, with solutions greater than 7–10 μmol being fatal to 100% of the animals.

In both of these experimental animals, injections of the aluminum containing solutions can be made into several different regions. Most researchers wishing to achieve generalized CNS effects inject the solutions into the cerebrospinal fluid, i.e., either into the lateral ventricles or cisterna magna. Such injections generally insure widespread CNS contamination with aluminum. For more localized effects, aluminum can be infused slowly into specific brain sites. Although such injections do reduce the generalized spread of aluminum, they do not preclude it.

As mentioned earlier, systemic injections of aluminum tartrate or lactate have been used in rabbits to produce an aluminum encephalopathy.[27] Apparently injections of 7.5 mg/kg/d or higher are required to produce an encephalopathy that is observable after approximately 20 d.

9.3. Human Research

The major methodological problems encountered at the human level involve assessment of aluminum levels rather than introduction of aluminum into the CNS. It is difficult to determine the level of CNS aluminum by peripheral aluminum assessment. As previously mentioned, peripheral blood aluminum levels reflect little more than temporary circulating peripheral levels. Even cerebrospinal fluid aluminum level does not correlate well with increased brain aluminum content.[1]

An additional difficulty involves the patchy distribution of aluminum. This makes an accurate assessment of aluminum content difficult without a large number of samples from various brain regions. This circumstance may, in part, account for the differences observed by different researchers. The reader is referred to Crapper McLachlan[23] for further discussion.

The other major methodological problem that may be responsible for different results involves the methods used to assay for aluminum. Since aluminum is ubiquitous in our environment, great care must be taken to prevent contamination. The specific techniques used for the quantitative assessment of aluminum were recently reviewed by King et al.[45] The reader is referred to this excellent review for further details on quantitative aluminum assay techniques.

9.4. Species Differences

As mentioned several times in this chapter, not all species respond in similar manner to elevated aluminum levels. Cats and rabbits have been the most frequently used experimental animals as they develop the characteristic neurofibrillary tangle and the progressive encephalopathy characteristic of aluminum exposure. In contrast to the cat and rabbit, two species of rats developed neither the progressive encephalopathy nor neurofibrillary tangles even at 10 times the dose effective in cats and rabbits.[44] The guinea pig and rhesus macaque monkey also develop tangles, but DeBoni et al.[27] report that the monkey does so only after a prolonged period of time, 380 d or more. Though certain human neurons, i.e., cerebral cortical neurons have been shown to be susceptible to in vitro inductions of neurofibrillary tangles, such tangles have not been observed in vivo in human dialysis patients despite very high tissue concentrations of aluminum.

10. Concluding Comments

We now have a fairly clear picture of the sequence of events involved in aluminum encephalopathy in the experimental animal. Aluminum appears to enter the brain and bind to chromatin and possibly to other cellular structures. It causes an increase in the production of protein and neurofibrillary tangles subsequently appear in the cytoplasm. Shortly after the appearance of the tangles subtle behavioral changes are observed. These are followed by electrophysiological, neurochemical, and anatomical changes that become progressively worse. It is difficult to determine which of the events are causally related, e.g., it is not known what role the tangles play in the series of cascading events, whether they are critical for the electrophysiological, behavioral, and anatomical changes, or whether some other effect of aluminum is responsible. Since the effects of aluminum are progressive, this suggests that the direct effect of competition with calcium is not likely to be the major mediating factor. Rather, cell changes secondary to possible nucleolar interferences or the accumulation of tangles would be more consistent with the progressive encephalopathy.

Some of the interest in aluminum has been based on its possible involvement in human disease processes, particularly dialysis dementia and Alzheimer's disease. Although aluminum's role in dialysis dementia is now well-established, it is equally clear that it is not the primary etiological agent in Alzheimer's disease, although it may play a secondary role. Unfortunately, we have a greater understanding of the effect of aluminum in nonhuman animals than we do in man. The metal has dramatic effects in certain species, with little or no effects on others. What we do not know is the effect it has on the human species, e.g., there is uncertainty as to whether it induces neurofibrillary tangles in human neurons. Animal research to date suggests that an encephalopathy is seen only in those species that develop tangles such that if humans do undergo an aluminum encephalopathy, but do not develop tangles as suggested, this will be the first species reported to do so. Perhaps there is a greater resistance in primates, since monkeys undergo an encephalopathy only after long periods of exposure, very different from the reaction seen in cats and rabbits. These results clearly point out large species differences in the reaction to aluminum, the exact basis of which would be of interest in further understanding its effects.

However, regardless of its effect in human disease or toxicological states, aluminum is an interesting neurobiological tool in its own right. The progressive accumulation of neurofibrillary tangles in the cytoplasm and dendrites makes the use of aluminum an interesting tool to study the role of neurotubules and filaments. A massive amount of research has been generated over the last decade to increase our understanding of neurotubules and neuroplasmic transport, and their role in cellular function. Though most other tubule disruptors are highly toxic to cells, the aluminum-induced tangle is slow, progressive, and compatible with longer cell life. Further, its dendritic, but not axonal involvement makes it a potentially invaluable tool to separate out other effects of the toxin.

Thus, we have clearly made large advances in understanding the effects of aluminum, and are now at a point that the research pace should quicken. Fortunately aluminum does not present an environmental hazard, but its potential role in human disease and use as a neurobiological tool warrant further research to understand the mechanisms of its actions.

Acknowledgments

This research has been supported by Grants #A0292 and G0165 from the Natural Science and Engineering Research Council of Canada.

References

1. Alfrey, A. C. (1980) Aluminum metabolism in uremia. *Neurotoxicol.* **1**, 43–53.
2. Alfrey, A. C., LeGendre, G. R. and Kaehny, W. D. (1976) The dialysis encephalopathy syndrome. Possible aluminum intoxication. *N. Eng. J. Med.* **294**, 184–188.
3. Anderton, B., Ayers, M. and Thorpe, R. (1978) Neurofilaments from mammalian central and peripheral nerve share certain peptides. *FEBS Lett.* **96**, 159–163.
4. Arieff, A. I., Cooper, J. D., Armstrong, D. and Lazarowitz, V. C. (1979) Dementia renal failure, and brain aluminum. *Ann. Intern. Med.* **90**, 741–747.
5. Ball, J. H., Butkus, D. E. and Madison, D. S. (1977) Effect of subtotal parathyroidectomy on dialysis dementia. *Nephron.* **18**, 151–155.
6. Berlyne, G. M. (1978) Plasmapharesis, aluminum, and dialysis dementia. *Lancet* **2**, 1155–1156.
7. Berlyne, G. M., Ben-Ari, J., Knopf, E., Yagil, R., Weinberger, G. and Danovitch, G. M. (1972) Aluminum toxicity in rats. *Lancet* **1**, 564–567.
8. Blaustein, M. P. and Goldman, D. E. (1968) The action of certain polyvalent cations on the voltage-clamped lobster axon. *J. Gen. Physiol.* **51**, 279–291.
9. Bowdler, N. C., Beasley, D. S., Fritze, C., Goulette, A. M., Hatton, J. D., Hession, J., Ostman, D. L., Rugg, D. J. and Schmittdiel, C. J. (1979)

Behavioral effects of aluminum ingestion on animal and human subjects. *Pharmacol. Biochem. Behav.* **10**, 505–512.

10. Cam, J. M., Luck, V. A., Eastwood, J. B. and DeWardner, H. E. (1976) The effect of aluminum chloride orally on calcium, phosphorus, and aluminum in normal subjects. *Clin. Sci. Mol. Med.* **51**, 407–414.

11. Campbell, I. R., Cass, J. S., Cholak, J. and Kehoe, R. A. (1957) Aluminum in the environment of man. A review of its hygenic status. *AMA Arch. Ind. Health.* **15**, 359–448.

12. Cartier, F., Allain, P., Gary, J., Chatel, M., Menault, F. and Pecker, S. (1978) Encephalopathic myclonique progressive des dialyses. Role de lean utilisee pour l'hemodialyse. *Nouv. Presse. Med.* **7**, 97–102.

13. Chokroverty, S., Bruetman, M. E., Berger, V. and Reyes, M. D. (1976) Progressive dialytic encephalopathy. *J. Neurol. Neurosurg. Psychiat.* **39**, 411–419.

14. Crapper, D. R. (1973) Experimental neurofibrillary degeneration and altered electrical activity. *Electroencephal. Clin. Neurophysiol.* **35**, 575–588.

15. Crapper, D. R. (1976) Functional consequences of neurofibrillary degeneration. In: R. D. Terry and S. Gershon (Eds.) *Neurobiology of Aging*, Raven, New York, pp. 405–432.

16. Crapper, D. R. and Dalton, A. J. (1973) Alterations in short-term retention conditioned avoidance response acquisition and motivation following aluminum induced neurofibrillary degeneration. *Physiol. Behav.* **10**, 925–933.

17. Crapper, D. R. and Dalton, A. J. (1973) Aluminum induced neurofibrillary degeneration, brain electrical activity, and alterations in acquisition and retention. *Physiol. Behav.* **10**, 935–945.

18. Crapper, D. R. and DeBoni, U. (1977) Aluminum and the genetic apparatus in Alzheimer Disease. In: K. Nandy and I. Sherwin (Eds.) *The Aging Brain and Senile Dementia.* Vol. 23 of Advances in Behavioral Biology. Plenum, New York, pp. 229–246.

19. Crapper, D. R. and DeBoni, U. (1978) Brain aging and Alzheimers disease. *Can. Psychiatr. Assoc. J.* **23**, 229–233.

20. Crapper McLachlan, D. R. and DeBoni, U. (1980) Aluminum in human brain disease—an overview. *Neurotoxicol.* **1**, 3–16.

21. Crapper, D. R., Krishnan, S. S. and Dalton, A. J. (1973) Brain aluminum distribution in Alzheimers disease and experimental neurofibrillary degeneration. *Science* **180**, 511–513.

22. Crapper, D. R., Krishnan S. S. and Quittkat, S. (1976) Aluminum neurofibrillary degeneration and Alzheimers disease. *Brain* **99**, 67–80.

23. Crapper McLachlan, D. R., Krishnan, S. S., Quittkat, S. and DeBoni, U. (1980) Brain aluminum in Alzheimer disease: influence of sample size and case selection. *Neurotoxicol.* **1**, 25–32.

24. Crapper, D. R. and Tomko, G. J. (1975) Neuronal correlates of an encephalopathy associated with aluminum neurofibrillary degeneration. *Brain Res.* **97**, 253–264.

25. Dahlstrom, A., Heiwall, P. O. and Larsson, P. A. (1975) Comparison between the effects of colchicine and lumicolchicine on axonal transport in rat motor neurons. *J. Neurol. Sci.* **37**, 305–311.
26. DeBoni, U. and Crapper, D. R. (1978) Paired helical filaments of the Alzheimer type in cultured neurons. *Nature* **271**, 566–568.
27. DeBoni, U., Otvos, A., Scott, J. W. and Crapper, D. R. (1976) Neurofibrillary degeneration induced by systemic aluminum. *Acta Neuropath.* **35**, 285–294.
28. DeBoni, U., Scott, J. W. and Crapper, D. R. (1974) Intracellular aluminum binding: A histochemical study. *Histochem.* **40**, 31–37.
29. DeBoni, U., Seger, M. and Crapper McLachlan, D. R. (1980) Functional consequences of chromatin bound aluminum in cultured human cells. *Neurotoxicol.* **1**, 65–81.
30. Dunea, G., Mahurkar, S. D., Mamdani, B. and Smith, E. C. (1978) Role of aluminum in dialysis dementia. *Ann. Int. Med.* **88**, 502–504.
31. Elliott, H. L., MacDougall, L. and Fell, G. S. (1978) Aluminum toxicity syndrome. *Lancet* **1**, 1203.
32. Elliott, H. L., MacDougall, A. I., Fell, G. S. and Gardiner, P. H. E. (1978) Plasmapharesis, aluminum and dialysis dementia, *Lancet* **2**, 1255.
33. Elliott, H. L., MacDougal, A. I., Haase, G., Cumming, R. L. C., Gardner, P. H. E. and Fell, G. S. (1978) Plasmopharesis in the treatment of dialysis encephalopathy. *Lancet* **2**, 940–941.
34. Embree, L. J., Hamberger, A. and Sjostrand, J. (1967) Quantitative cytochemical studies and histochemistry in experimental neurofibrillary degeneration. *J. Neuropath. Exp. Neurol.* **26**, 427–436.
35. Galle, P., Berry, J. P. and Duckett, S. (1980) Electron microprobe ultrastructural localization of aluminum in the rat brain. *Acta Neuropathol.* **49**, 245–247.
36. Gorsky, J. E., Dietz, A. A., Spencer, H., and Osis, D. (1979) Metabolic balance of aluminum. *Clin. Chem.* **25**, 1739–1743.
37. Gross, G. W. (1975) The microstream concept of axoplasmic and dendritic transport. In: G. W. Kruetzberg (Ed.), *Advances in Neurology*, Vol. 12, Raven, New York, pp. 283–296.
38. Harrison, W. H., Codd, E. and Gray, R. M. (1972) Aluminum inhibition of hexokinase. *Lancet* **2**, 277.
39. Hetnarski, B., Wisniewski, H. M., Iqbal, K., Dziedzic, J. D. and Lajtha, A. (1980) Central cholinergic activity in aluminum-induced neurofibrillary degeneration. *Ann. Neurol.* **7**, 489–490.
40. Kaehny, W. D., Alfrey, A. C., Holman, R. E. and Shorr, W. J. (1977) Aluminum transfer during hemodialysis. *Kidney Int.* **12**, 361–365.
41. Kaehny, W. D., Hegg, A. P. and Alfrey, A. C. (1977) Gastrointestinal absorption of aluminum containing antacids. *N. Eng. J. Med.* **296**, 1389–1390.
42. Karlik, S. J., Eichorn, G. L. and Crapper McLachlan, D. R. (1980) Molecular interactions of aluminum with DNA. *Neurotoxicol.* **1**, 83–88.
43. Karlik, S. J., Eichorn, G. L., Lewis, P. N. and Crapper, D. R. (1979) Aluminum interactions with D.N.A. Abstract: *XIth Int. Cong. Biochem.* Toronto.

44. King, G. A., DeBoni, U. and Crapper, D. R. (1975) Effect of aluminum upon conditioned avoidance response acquisition in the absence of neurofibrillary degeneration. *Pharmacol. Biochem. Behav.* **3**, 1003–1009.

45. King, S. W., Savory, J. and Wills, M. R. (1981) The clinical biochemistry of aluminum. *CRC Crit. Rev. Clin. Lab. Sciences.* **14**, 1–20.

46. King, S. W., Wills, W. R. and Savory, J. (1979) Serum binding of aluminum. *Res. Commun. Chem. Pathol. Pharmacol.* **26**, 161–169.

47. Klatzo, I., Wisniewski, H. and Streicher, E. (1965) Experimental production of neurofibrillary degeneration. I. Light microscopic observations. *J. Neuropath. Exp. Neurol.* **24**, 187–197.

48. Kovalchik, M. T., Kaehny, W. D., Jackson, T. and Alfrey, A. C. (1978) Aluminum kinetics during hemodialysis. *J. Lab. Clin. Med.* **92**, 712–720.

49. Krishnan, S. S., Gillespie, K. A. and Crapper, D. R. (1972) Determination of aluminum in biological material by atomic absorption spectrophotometry. *Anal. Chem.* **44**, 1469–1470.

50. Kushelevsky, A., Yagil, R., Alfasi, Z. and Berlyne, G. M. (1976) Uptake of aluminum ion by the liver. *Biomedicine* **25**, 59–60.

51. Lai, J. C., Guest, J. F., Leung, T. K. C., Lim, L. and Davidson, A. N. (1980) The effects of cadmium, manganese, and aluminum on sodium-potasium-activated and magnesium-activated adenosine triphosphatase activity and choline uptake in rat brain synaptosomes. *Biochem. Pharmacol.* **29**, 141–146.

52. Liss, L., Ebner, K., Couri, D. and Cho, N. (1975) Alzheimer disease. A possible animal model. Abstract: *IV Panamerican Congress of Neurology* Mexico, 1975.

53. Liwnicz, B. H., Kristensson, K., Wisniewski, H. M., Shelanski, M. L. and Terry, R. D. (1974) Observations on axoplasmic transport in rabbits with aluminum induced neurofibrillary tangles. *Brain Res.* **80**, 413–420.

54. Lundin, A. P., Caruso, C., Sass, M. and Berlyne, G. M. (1978) Ultrafilterable aluminum in serum of normal man. *Clin. Res.* **26**, 636A.

55. Masselot, J. P., Adhemar, J. P., Jaudon, M. C., Kleinknecht, D. and Galli, A. (1978) Reversible dialysis encephalopathy: Role for aluminum containing gels. *Lancet* **2**, 1386–1387.

56. Mayor, G. H., Keiser, J. A., Makdani, D. and Ku, P. K. (1977) Aluminum absorption and distribution: effect of parathyroid hormone. *Science*, **197**, 1187–1189.

57. Mayor, G. H., Remedi, R. F., Sprague, S. M. and Lovell, K. L. (1980) Central nervous system manifestations of oral aluminum: effect of parathyroid hormone. *Neurotoxicol.* **1**, 33–42.

58. McDermott, J. R., Smith, A. I., Iqbal, K. and Wisniewski, H. M. (1977) Aluminum and Alzheimers disease. *Lancet* **2**, 710–711.

59. McDermott, J. R., Smith, A. I., Iqbal, K. and Wisniewski, H. M. (1979) Brain aluminum in aging and Alzheimer disease. *Neurol.* **29**, 809–814.

60. McDermott, J. R., Smith, A. I., Ward, M. K., Parkinson, I. S. and Kerr, D. N. S. (1978) Brain aluminum concentration in dialysis encephalopathy. *Lancet* **1**, 901–903.
61. McLaughlin, A. I. G., Kazantizis, G., King, E., Teare, D., Porter, R. J. and Owen, R. (1962) Pulmonary fibrosis and encephalopathy associated with the inhalation of aluminum dust. *Br. J. Ind. Med.* **19**, 253–263.
62. Miller, C. A. and Levine, E. M. (1974) Effect of aluminum gels on cultured neuroblastoma cells. *J. Neurochem.* **22**, 751–758.
63. Potocka, J. (1971) The influence of Al in cholinesterase and acetylcholinesterase activity. *Acta Biologica Medica Germinica* **26**, 845–846.
64. Perl, D. P. and Brody, A. R. (1980) Alzheimers disease: x-ray spectrometric evidence of aluminum accumulation in neurofibrillary tangle-bearing neurons. *Science* **208**, 297–299.
65. Perl, D. P. and Brody, A. R. (1980) Detection of aluminum by SEM x-ray spectrometry within neurofibrillary tangle-bearing neurons of Alzheimers disease. *Neurotoxicol.* **1**, 133–137.
66. Petit, T. L. (1982) Neuroanatomical and clinical neuropsychological changes in aging and senile dementia. In: F. I. M. Craik and S. Trehub (Eds.) *Aging and Cognitive Processes*, Plenum, New York, pp. 1–25.
67. Petit, T. L., Biederman, G. B. and McMullen, P. A. (1980) Neurofibrillary degeneration, dendritic dying back, and learning-memory deficits after aluminum administration: Implications for brain aging. *Exp. Neurol.* **67**, 152–162.
68. Petit, T. L. and Isaacson, R. L. (1977) Deficient brain development following colcemid treatment in postnatal rats. *Brain Res.* **132**, 380–385.
69. Poisson, M., Mashaly, R. and Labkiri, B. (1978) Dialysis encephalopathy: Recovery after interruption of aluminum intake. *Br. Med. J.* **2**, 1610–1611.
70. Rozas, V. V., Port, F. K. and Rutt, W. M. (1978) Progressive dialysis encephalopathy from dialysis aluminum. *Arch. Int. Med.* **138**, 1375–1377.
71. Schlaepfer, W. and Freeman, L. (1978) Neurofilament proteins of rat peripheral nerve and spinal cord. *J. Cell Biol.* **78**, 653–662.
72. Schreeder, M. T. (1979) Dialysis encephalopathy. *Arch. Int. Med.* **139**, 510–511.
73. Schubert, P. and Kreutzberg, G. W. (1975) Parameters of dendritic transport. In: G. Kreutzberg (Ed.) *Advances in Neurology*, Vol. 12, Raven, New York, pp. 255–268.
74. Schubert, P., Kreutzberg, G. W. and Lux, H. D. (1972) Neuroplasmic transport in dendrites: Effects of colchicine on morphology and physiology of motoneurons in the cat. *Brain Res.* **47**, 331–343.
75. Sears, W. G. and Eales, L. (1973) Aluminum-induced porphyria in the rat. *IRCS (Med. Sciences)* **1**, 13.

76. Selkoe, D. J., Liem, R. K. H., Yen, S. H. and Shelanski, M. L. (1979) Biochemical and immunological characterization of neurofilaments in
experimental neurofibrillary degeneration induced by aluminum.
Brain Res. **163**, 235–252.

77. Siebert, F. and Wells, H. (1929) The effect of aluminum on mammalian blood and tissues. *Arch. Pathol.* **8**, 230–261.

78. Sjostrand, J. and Hansson, H. (1971) Effect of colchicine on the transport of axonal protein in the retinal ganglion cells of the rat. *Exp.
Eye Res.* **12**, 261–269.

79. Still, C. N. and Kelley, P. (1980) On the incidence of primary degenerative dementia vs. water fluoride content in South Carolina.
Neurotoxicol. **1**, 125–131.

80. Terry, R. D. and Pena, C. (1965) Experimental production of
neurofibrillary degeneration. (2) Electron microscopy phosphatase
histochemistry and electron probe analysis. *J. Neuropath. Exp.
Neurol.* **24**, 200–210.

81. Tipton, I. H., Cook, M. J., Foland, J. M., Rittner, J., Hardwich, M. and
McDaniel, K. K. (1959) Oak Ridge Nat. Lab. Rep. Central File No.
59-8-106, Oak Ridge, Tennessee.

82. Trapp, G. A. (1980) Studies of aluminum interaction with enzymes
and proteins—the inhibition of hexokinase. *Neurotoxicol.* **1**, 89–100.

83. Trapp, G. A., Miner, G. D., Zimmerman, R. L., Mastri, A. R. and
Heston, L. L. (1978) Aluminum levels in brain in Alzheimers Disease.
Biol. Psychiat. **13**, 709–718.

84. Ulmer, D. D. (1976) Toxicity from aluminum antacids. *N. Eng. J. Med.*
294, 218–219.

85. Wallis, C. and Melnick, J. L. (1967) Concentration of viruses on aluminum and calcium salts. *Am. J. Epidem.* **85**, 459–468.

86. Wenk, G. L. and Stemmer, K. L. (1981) The influence of ingested aluminum upon norpinephrine and dopamine levels in the rat brain.
Neurotoxicol. **2**, 347–353.

87. Westrum, L. E., White, L. and Ward, A. (1964) Morphology of the experimental epileptic focus. *J. Neurosurg.* **21**, 1033–1046.

88. Wisniewski, H. M., Iqbal, K. and McDermott, J. R. (1980) Aluminum-
induced neurofibrillary changes: its relationship to senile dementia
of the Alzheimers type. *Neurotoxicol.* **1**, 121–124.

89. Wisniewski, H., Narkiewicz, O. and Wisniewska, K. (1967) Topography and dynamics of neurofibrillar degeneration in aluminum encephalopathy. *Acta Neuropathol.* **9**, 127–133.

90. Wisniewski, H., Terry, D. and Hirano, A. (1970) Neurofibrillary pathology. *J. Neuropath. Exp. Neurol.* **29**, 163–176.

91. Womak, F. C. and Colwick, S. P. (1979) Proton-dependent inhibition
of yeast and brain hexokinase by aluminum in ATP preparations.
Proc. Natl. Acad. Sci. USA **76**, 5080.

92. Yamada, K. M., Spooner, B. S. and Wessells, N. K. (1970) Axon growth:
roles of microfilaments and microtubules. *Proc. Natl. Acad. Sci. USA*
66, 1206–1209.

Chapter 9

Lithium Effects on Bipolar (Manic–Depressive) Illness and Other Behavior

Michael H. Sheard

1. Introduction

Lithium's use in medicine dates back to the 1850s and the notion that it might be useful for the treatment of gout because lithium urate was the most soluble urate in the test tube. This latter fact underlay the serendipitous discovery by Cade in 1949[4] that lithium could be used to treat pathological excited states in humans. The subsequent development of the evidence to support this original finding has been well-documented in a number of review articles by Schou,[30] Prien,[25] and Gerbino et al.[12] In summary, lithium was first found to be efficacious for the treatment of acute states of mania and later of value in the prophylaxis of both the manic and depressive episodes of bipolar (manic–depressive) illness. More recently, evidence has accumulated to demonstrate that lithium can be effective in the treatment or prophylaxis of acute episodes of unipolar depression. Thus lithium emerges as a unique agent able to control two abnormal states of behavior with apparently oppo-

Michael H. Sheard: Yale University School of Medicine, Department of Psychiatry, Connecticut Mental Health Center, New Haven, Connecticut

site clinical characteristics, one with excess activation and one with decreased activation.

2. The Specificity of Lithium

At first sight it appears as if lithium is specific for mood disorders characterized by either abnormal exhilaration or depression. Indeed some clinicians would argue that lithium responsiveness is evidence for a mood disorder.[45] There is, however, mounting evidence to suggest that such specificity is relative and that lithium might have other properties in addition to mood stabilization. Evidence of neuroleptic properties for lithium has come from studies in schizoaffective and schizophrenic disorders. For example, Prien et al.[26] reported no difference between lithium and chlorpromazine in treating mildly active schizoaffective patients. Small et al.[43] used a crossover design with lithium alternating with placebo for 4-week periods in 22 chronic schizophrenic patients who were maintained on their chronic neuroleptic medication. There is no question here about the absence of a mood disturbance. They found significantly less pathology during lithium treated periods. Biederman et al.[2] compared the effects of lithium plus haloperidol with placebo plus haloperidol in excited schizoaffective patients. They found statistically significant differences in favor of lithium plus haloperidol and claimed no specificity of lithium for manic symptoms. However, they did report the fact that the "affective" schizoaffectives showed more improvement on the lithium/haloperidol combination than the "schizophrenic" schizoaffectives. In summary, lithium appears to affect favorably the course of some schizophrenic patients particularly when used in combination with other neuroleptics.

In addition to schizophrenia there are a variety of other episodic or periodic psychopathological states for which lithium responsiveness has been claimed. These have been reviewed by Kline and Simpson[18] and Gerbino et al.[12] and include some cases of epilepsy, unstable character disorder,[29] periodic catatonia, cycloid psychosis,[24] and episodic alcoholism associated with depression.[27,28]

3. Lithium and Aggressive Behavior

The best evidence for a lack of specificity comes from studies that have been performed on the effect of lithium on aggressive behavior. Weischer[49] studied the effect of putting lithium into the water of Siamese fighting fish and into the drinking water of male rodents (mice and hamsters). A significant reduction in aggressive behavior was found with large interindividual variations in serum and brain lithium levels. Similar experiments performed by this author corroborated these findings with isolated mice with the additional finding that there was no reduction in overall motor activity. Lithium was also shown to inhibit defensive aggressive behavior in rats as measured by footshock elicited fighting at doses that did not alter the flinch or jump threshold for individual rats.[9,35] Male rats habituated to home cages are naturally very aggressive towards male intruders and were shown to lose this aggressiveness when given lithium in their drinking water.[37] These forms of aggressive behavior can be categorized as more or less naturalistic in character. A more "pathological" form of aggressive behavior is that seen when brain serotonin is lowered, for example, by p-chlorophenylalanine (PCPA). PCPA is a compound that depletes brain serotonin (5-HT) by inhibiting the enzyme tryptophan hydroxylase involved in the synthesis of 5-HT. Rats treated with PCPA show a remarkable increase in both aggressive and sexual behavior.[34] Pretreatment with lithium for 5 d prior to the administration of PCPA blocked the abnormal increase in both aggressive and sexual behavior.[36]

These findings demonstrate that lithium can inhibit aggressive behavior in a wide variety of animal models. Clinical observations in humans, such as those of Gershon[13] in epileptics, showing a reduction in disturbed interseizure behavior characterized by impulsivity, aggressiveness, and excitement, and Dostal and Zvolsky[8] in disturbed mental retardates, showing a reduction in aggressive combative behavior with lithium, furthered the notion that lithium might have a specific antiaggressive action. However, in these instances the attribution of such an action is complicated by the use of other medications and the presence of additional deficits. To provide further evidence for lithium's antiaggressive action, a series of clinical trials were performed in prison settings. Male subjects were chosen on the basis of a prior history of violent assaultive crime and continuing aggressive behavior in prison. Psychosis, brain damage, and mental retardation were excluded.

In the first study male subjects ages 21–43 were given either lithium or placebo for alternating 4 week periods.[38] The serum lithium was maintained between 0.6 and 1.5 mEq/L by adjustment of the daily dose. During the lithium periods, there was a significant reduction in aggressive affect (self-rated items of anger and tension) as well as a reduction in the number of disciplinary tickets received for physical or verbally threatening aggressive behavior compared with placebo periods. This study was single-blind; another single-blind study was performed by Tupin et al.[46] on a prison population of 27 chronically aggressive individuals. These subjects were treated for up to 1.5 yr. Twenty-one subjects improved on lithium; however, they had mixed diagnoses, including schizophrenia and brain damage. Sheard's[39] second study was also single-blind and tested the effect of lithium in a younger age group of male delinquents aged 16–23. The subjects were incarcerated for the first 4 months of this trial and then were out on parole from 1–12 months. The clinical characteristics were best described as impulsive aggressive behavior disorders without evidence of psychosis or brain damage and with IQs over 85. The results of this trial also revealed fewer aggressive episodes when serum lithium was between 0.6 and 1.2 mEq/L than between 0.0 and 0.6 mEq/L. These suggestive results led to a full-scale double-blind study[40] in 16–24-yr-old male inmate volunteers with the characteristics described above. The experimental design consisted in a 1 month predrug and postdrug observation with an invervening 3 months on either lithium or placebo. The results showed a statistically significant reduction in serious aggressive episodes for the lithium treatment group. This evidence, together with several impressive single case studies reported by Baastrup,[1] Forssman and Walinder,[11] Shader et al.,[33] Lion,[19] Kerr,[17] and Panter[23] is strongly in favor of lithium having an antiaggressive action. The details of these studies have been reviewed by Sheard and Marini.[41]

4. Mechanism of Lithium's Action on Behavior

Before turning to an examination of specific neurobiological actions that might explain lithium's action on beahvior, it is necessary to consider the possibility of more general actions. These include toxic, motor, sensory, reaction time coordination, cognitive, and placebo response. Though toxicity would be an explanation for reduced aggression in some studies,[8] there was no evidence of toxicity in the prisoner studies[38–40] and thus this explanation can be ruled out. It still is a possibility that a slight general malaise short of toxicity might cause a reduction in aggression. However, against this it can be argued that manipulations causing more obvious malaise do not necessarily reduce aggression in animal models and may indeed enhance aggression sometimes. Moreover in the double-blind prison study, the time of maximum lithium effect, i.e., during the third month, when no aggressive outbursts occurred, coincided with the time of fewest complaints on a symptom checklist. Tests that have been used to exclude a direct effect on motor strength include right- and left-hand grip and a swim to exhaustion test.[44] Motor and eye–hand coordination has been tested with the pursuit rotor, a turning block, and forced tapping tests. None of these tests differentiated lithium from placebo subjects. Lastly, in the double-blind prisoner study, the reduction in violent aggressive behavior was not correlated with a reduction in other forms of antisocial behavior resulting in minor infractions. Johnson[15] has postulated that changes in sensory reactivity either in the direction of hyper- or hypoalgesia could be part of lithium's action on behavior. However, experiments on aggressive behavior in animal models using mice and rats[21,35] have not revealed significant sensory changes and symptoms of hypoalgesia or hyperalgesia have not been reported as side effects in man.[47]

Schou[30] has reported a reduced choice reaction time, but in our study[40] no significant differences were found in choice reaction time, or informed delay reaction time, between placebo and lithium subjects. Thus, change in reaction time could not account for the reduction in aggressive behavior in this study.

An impairment in cognitive functioning has been reported at higher dose levels of lithium by Judd et al.[16] and Demers and Heninger[5] did find a significant difference on the digit symbol substitution test (DSST) comparing lithium-treated subjects with matched normal controls. However, at the serum levels achieved in the double-blind prison study,[40] while there was a tendency to lower performance on the DSST, this did not attain statistical significance. It is possible that an action on cognitive functioning is an important mechanism of action for lithium. This is demonstrated by some anecdotal statements from patients "I am more amenable to suggestion . . . I get angry but I don't lose control of myself . . . [33]" "I stop to think before mouthing off at the guards.[40]" Finally a placebo-type of response has to be considered and since lithium has a rather specific set of side effects absolute double-blind control is difficult to achieve. However, Schou et al.[31] and Schou[32] have reported that normal volunteers failed to distinguish lithium from placebo and in the double-blind prisoner study placebo itself did not produce any significant inhibition. These facts suggest that a placebo response cannot explain lithium's antiaggressive action.

5. Neurobiological Actions of Lithium

Lithium has many biological actions that make its relatively specific action on behavior all the more puzzling. Although it possesses the smallest radius of all alkali metals, in physiological solution it is readily hydrated and comes to resemble calcium and magnesium. Lithium can replace sodium and maintain the resting and action potentials of nerve cells. Lithium has widepsread effects on enzyme systems; in particular, adenylate cyclase activity in many cellular systems is inhibited by lithium and this helps to explain many of its endocrine effects leading to profound effects on many metabolic systems. Many of the biological effects ascribed to lithium have resulted from the use of acute doses while the clinical effects result from chronic administration. It may well be that changes in receptor function, perhaps through lithium's action on cyclic AMP or via a conformational change in the receptors, may in the long run be more important determinants affecting neural transmission.

6. Theory of Lithium's Action

The evidence suggests that although lithium may not be completely specific for mania, at least it has a relative specificity for mood disorders and effects on pathological anger must now be added to effects on pathological euphoria and depression. The unique circumstances of a simple metallic ion with dramatic effect upon a major type of mental illness stimulates great interest and research activity. As Schou[30] has pointed out, animal models for the endogenous psychoses do not exist. However, certain features of psychosis, such as hyperactivity, irritability, increased aggression, and even "depression" may be more readily obtainable in animals models than others such as elation.

Some clinicians have pointed out that irritability and anger are as characteristic of manic attacks as elation and frequently there is a very dysphoric state with negative emotions predominating. Thus it is possible that a study of lithium's action in certain animal models of aggression may help to shed light on lithium's mechanism of action in manic depressive illness. Lyttkens[20] has suggested previously that aggressiveness might characterize lithium-responsive conditions. The brain adds an affective coloration to stimuli in a negative or positive dimension. Bipolar patients can be said to have a tendency to overreact to both negative and positive stimuli. The overreaction to negative stimuli leads to depression and the overreaction to positive stimuli leads to mania. Abnormal affects, such as brief spells of depression, euphoria, or fear, are seen in certain seizure disturbances that involve the temporal lobes or with electrical stimulation of these regions. These facts suggest that affective valence is added to the internal processing of stimuli from such limbic sites as the amygdala and hippocampus. These sites are known to receive an important 5-HT input.

It has been noted above that rats treated with PCPA, which depletes brain 5-HT, show hyperaggressiveness, hypersexuality, increased motor activity in a novel environment and enhanced reactivity to novel auditory stimuli. This syndrome shows some of the features of a manic state and, as we have shown, all these symptoms in the animal model can be blocked by lithium.[36] The pathological symptoms released by depletion of brain 5-HT suggest that 5-HT normally exerts a tonic inhibitory control in the forebrain. If this were the case then agents which increased 5-HT release could be expected to enhance the inhibitory control. Sheard and Aghajanian[42] showed that rats pretreated with lithium showed increased 5-HT metabolism in brain particularly when the 5-HT neuronal systems were electrically stimulated. It was in fact this finding that led us to the trials of lithium in animal models of aggression.

Further evidence implicating 5-HT in lithium's mechanism of action in mood disorders comes from clinical observations during the use of d-lysergic acid diethylamide. This hallucinogenic compound, which is a 5-HT agonist, can induce amplifications of affect, such as fear, euphoria, depression, and precipitate psychosis with strong affective components. These drug-induced psychoses have also been reported to respond to lithium.[14] In human subjects, it has been well-documented that manic episodes may be precipitated by antidepressant drugs such as tricyclic compounds or monoamine oxidase inhibitors (Bunney[3]). Moreover, the chronic administration of antidepressant drugs to rats induces an increase in aggressive behaviors as reported by Eichelman and Barchas[10] and Mogilnicka and Przewlocka.[22] Generally the clinically effective antidepressant drugs have been found to enhance responses to iontophoretically applied 5-HT, for example, in the hippocampus[6] and amygdala.[48]

We have shown that the increase in aggressiveness with chronic antidepressant drug administration can be blocked by lithium. Thus it may be that 5-HT plays a role in inhibiting both the

exaggerated response to negative stimuli that is depression and
the exaggerated responses to positive stimuli that is mania, or exaggerated angry responses. Lithium may then be acting to enhance the responsivitity to or the functional efficacy of 5-HT. Some
support for this notion comes from a recent study by deMontigny
et al.,[7] who showed that some patients with depression who were
resistant to antidepressant drugs responded quickly when lithium
was added.

References

1. Baastrup, P. C. (1969) Practical clinical viewpoints regarding treatment with lithium, *Acta Psychiat. Scand. Suppl.* **207**, 12–18.
2. Biederman, J., Lerner, Y. and Belmaker, R. (1979) Combination of
   lithium plus haloperidol in schizoaffective disorders: a controlled
   study, *Arch. Gen. Psychiat.* **36**, 327–333.
3. Bunney, W. E., Jr. (1978) Psychopharmacology of the switch process
   in affective illness. In: M. A. Lipton, A. DiMascio and K. F. Killam
   (Eds.), Raven Press, New York, pp. 1249–1259.
4. Cade, J. F. J. (1949) Lithium salts in the treatment of psychotic excitement, *Med. J. Aust.* **36**, 349–352.
5. Demers, R. G. and Heninger, G. R. (1971) Visual-motor performance
   during lithium treatment: a preliminary report, *J. Clin. Pharmacol.*
   11, 274–279.
6. deMontigny, C. and Aghajanian, G. K. (1978) Tricyclic
   antidepressants: long term treatment increases responsivity of rat
   forebrain neurons to serotonin, *Science* **202**, 1303–1305.
7. deMontigny, C., Grunberg, F., Mayer, A., and Deschenes, J. P. (1981)
   Lithium induces rapid relief of depression in tricyclic
   antidepressant drug non responders, *Brit. J. Psychiat.* **138**, 252–256.
8. Dostal, T. and Zvolsky, P. (1970) Antiaggressive effect of lithium salts
   in several mentally retarded adolescents, *Int. Pharmacopsychiat.* **5**,
   203–207.
9. Eichelman, B., Thoa, N. B. and Perez-Cruet, J. (1973) Alkali metal cations: effects on aggression and adrenal enzymes, *Pharmacol.
   Biochem. Behav.* **1**, 121–123.
10. Eichelman, B. and Barchas, J. (1975) Facilitated shock-induced ag-
    gression following antidepressive medication in the rat, *Pharmacol.
    Biochem. Behav.* **3**, 601–606.
11. Forssman, M. and Walinder, J. (1969) Lithium treatment on atypical
    indication, *Acta Psychiat. Scand. Suppl.* **207**, 34–40.
12. Gerbino, L., Oleshansky, M. and Gershon, S. (1978) Clinical use and
    mode of action of lithium. In: M. A. Lipton, A. DiMascio, and K. F.
    Killam (Eds.), *Psychopharmacology: A Generation of Progress* Raven
    Press, New York, pp. 1261–1275.

13. Gershon, E. S. (1968) The use of lithium salts in psychiatric disorders, *Dis. Nerv. Syst.* **29**, 51–62.
14. Horowitz, H. A. (1975) The use of lithium in the treatment of drug induced psychotic reaction, *Dis. Nerv. Syst.* **37**, 159–163.
15. Johnson, F. N. (1975) Behavioral and cognitive effects of lithium: mechanisms. In: E. N. Johnson (Ed.), *Lithium Research and Therapy,* Academic Press, New York, pp. 315–335.
16. Judd, L. L., Hubbard, R. B., Janowksy, D. S. and Huey, L. (1976) Lithium induced mood and affect changes in normals, *Proc. Amer. Psychiat. Assoc.* (New Research Abst), p. 9.
17. Kerr, W. C. (1976) Lithium salts in the management of a child batterer, *Med. J. Aust.* **2**, 414–415.
18. Kline, N. and Simpson, G. (1975) Lithium in the treatment of conditions other than the affective disorders. In: F. N. Johnson (Ed.) *Lithium Research and Therapy,* Academic Press, New York. pp. 85–95.
19. Lion, J. R., Hill, J. and Madden, D. J. (1975) Lithium carbonate and aggression: a case report, *Dis. Nerv. Syst.* **36**, 97–98.
20. Lyttkens, G. (1969) Discussion, *Acta Psychiat. Scand. Suppl.* **207**, 25.
21. Mannisto, P. T. and Saarnivarra, L. (1972) Effect of lithium on the analgesia caused by morphine and two antidepressants in mice, *Pharmacology* **8**, 284–285.
22. Mogilnicka, E. and Przewlocka, B. (1981) Facilitated shock-induced aggression after chronic treatment with antidepressant drugs in the rat, *Pharmacol. Biochem. Behav.* **14**, 129–133.
23. Panter, B. M. (1977) Lithium in the treatment of a child abuser, *Am. J. Psychiat.* **134**, 1436–1437.
24. Perris, C. (1978) Morbidity suppressive effect of lithium carbonate in cycloid psychosis, *Arch. Gen. Psychiat.* **35**, 328–331.
25. Prien, R. F. (1975) The clinical effectiveness of lithium: comparisons with other drugs. In: F. N. Johnson (Ed.) *Lithium Research and Therapy,* Academic Press, New York, pp. 99–112.
26. Prien, R. F., McCaffey, E. and Klett, C. J. (1972) A comparison of lithium carbonate and chlorpromazine in the treatment of excited schizoaffectives, *Arch. Gen. Psychiat.* **27**, 182–189.
27. Reilly, E., Halmi, K. and Noyes, R. (1973) Electroencephalographic responses to lithium, *Int. Pharmacopsychiat.* **8**, 208–213.
28. Reynolds, C. M., Merry, J. and Coppen, A. (1977) Prophylactic treatment of alcoholism by lithium carbonate—an initial report, *Alcoholism* **1**, 109–111.
29. Rifkin, A., Quitkin, F., Carillo, C., Blumberg, A. and Klein, D. F. (1972) Lithium carbonate in emotionally unstable character disorder, *Arch. Gen. Psychiat.* **27**, 519–523.
30. Schou, M. (1968) Lithium in psychiatric therapy and prophylaxis, *J. Psychiat. Res.* **6**, 67–95.
31. Schou, M., Amdisen, A. and Thomsen, K. (1968) The effect of lithium on the normal mind. In: P. Baudis, E. Peterova and V. Sedevic (Eds.), *Psychiatric Progrediens,* Plzen, pp. 712–721.

32. Schou, M. (1976) Pharmacology and toxicology of lithium, *Ann. Rev. Pharmacol. Toxicol.* **16**, 231–243.

33. Shader, R. I., Jackson, A. H. and Dodes, L. M. (1974) The antiaggressive effects of lithium in man, *Psychopharmacology* **40**, 17–24.

34. Sheard, M. H. (1969) The effect of p-chlorophenylalanine on behavior in rats, relation to brain serotonin and 5-hydroxyindole acetic acid, *Brain Res.* **15**, 524–528.

35. Sheard, M. H. (1970) Effect of lithium on foot-shock aggression in rats, *Nature* **228**, 284–285.

36. Sheard, M. H. (1970) Behavioral effects of *p*-chlorophenylalanine in rats: inhibition by lithium, *Commun. Behav. Biol.* **5**, 71–73.

37. Sheard, M. H. (1973) Aggressive behavior; modification by amphetamine, p-chlorophenylalanine and lithium in rats, *Aggressologie* **14**, 323–326.

38. Sheard, M. H. (1971) Effect of lithium on human aggression, *Nature* **230**, 113–114.

39. Sheard, M. H. (1975) Lithium in the treatment of aggression, *J. Nerv. Ment. Dis.* **160**, 108–118.

40. Sheard, M. H., Marini, J. L., Bridges, C. I. and Wagner, E. (1976) The effect of lithium on impulsive-aggressive behavior in man, *Amer. J. Psychiat.* **133**, 1409–1413.

41. Sheard, M. H. and Marini, J. L. (1979) Lithium and aggressive behavior. In: S. Gershon, N. Kline, and M. Schou (Eds.) *Lithium: Controversies and Unresolved Issues*, Excerpta Medica, pp. 136–147.

42. Sheard, M. H. and Aghajanian, G. K. (1970) Neuronally activated metabolism of brain serotonin: effect of lithium, *Life Sci.* **9**, 285–290.

43. Small, J., Kellams, J., Milstein, V., and Moore, F. (1975) A placebo controlled study of lithium combined with neuroleptics in chronic schizophrenic patients, *Am. J. Psychiat.* **132**, 1315–1317.

44. Smith, D. F. and Smith, H. B. (1973) The effect of prolonged lithium administration on activity, reactivity and endurance in the rat, *Psychopharmacology* **30**, 83–88.

45. Taylor, M. and Abrams, R. (1973) The phenomenology of mania, *Arch. Gen. Psychiat.* **29**, 520–522.

46. Tupin, J. P., Smith, D. B., Classon, T. L., Kim, L. I., Nugent, A. and Groupe, A. (1973) Long-term use of lithium in aggressive prisoners, *Comprehensive Psychiat.* **14**, 311–317.

47. Vacaflor, L. (1975) Lithium side effects and toxicity: the clinical picture. In: F. N. Johnson (Ed.), *Lithium Research and Therapy*, Academic Press, London, pp. 211–226.

48. Wang, R. Y. and Aghajanian, G. K. (1979) Chronic treatment of rats with tricyclic antidepressants increases responsiveness of amygdaloid neurons to serotonin and norepinephrine, *Soc. Neurosci. Abst.* **5**, 356.

49. Weischer, M. L. (1969) Uber die antiaggressive wirkung von lithium, *Psychopharmacology* **15**, 245–254.

Trace Elements in Neurobiology-Conclusions

R. M. Smith and I. E. Dreosti

The past decade has seen a very rapid expansion of research endeavor in neurobiology that has accompanied the appearance of new concepts and new technologies aimed directly at this very complex field. The state of knowledge, however, is still developing and it is thus possible to expect a work such as the present one to present a reasonably comprehensive view of the emerging field of the neurobiology of the trace elements. Because they are nondegradable and both physically and chemically unique, the trace elements are intrinsically easy to quantify precisely and in very small amounts. Although it has not emerged in the present pair of volumes, the use of the electron probe to localize metabolic elements in single cells at the level of resolution of the electron microscope appears to present an opportunity to match, for trace elements, the specificity that immunocytochemistry has given to the location and mapping of neurotransmitter pathways in the central nervous system. The latter technique itself has been successfully used to locate two copper enzymes—dopamine beta-hydroxylase and cytochrome oxidase—in the brain and applications to the other trace elements will no doubt follow.

The most fruitful method of exploring trace element functions in the brain has been to induce a dietary deficiency of the element and to examine the distortion produced in functions of one kind or another. In some cases valuable additional information is obtained from a study of mutants whose genetic disability is related to a trace element function. Such mutants include the mottled mouse, whose impaired copper metabolism constitutes a valuable model of Menkes' disease in humans, strains of mice that are genetically resistant to cadmium toxicity and, in humans, the zinc-responsive disorder, acrodermatitis enteropathica. Insights from studies of these disorders have proved to be of general and lasting value.

A further relatively recent addition to the repertoire of the neurobiologist has been the use of behavioral studies, especially in humans, but also in animals. Although fraught with pitfalls relating to interpretation—a behavioral change can result from a bunion—such studies may nevertheless be exquisitely sensitive indicators of fine changes in brain performance. They are potentially most useful in detecting the early consequences of ingestion of materials harmful to the nervous system. The excellent chapter by B. Weiss, Metals and Behavioral Toxicology, explores the potentials and limitations of this emerging discipline.

The epidemiology of the effects of environmental disasters has also emerged as a powerful means of assessment of the effects of toxic metals in the human organism, especially of mercury. The epidemiology of lead toxicity has also been of value in understanding the limits of safety of this serious environmental pollutant.

In the field of toxicology, cell and organ culture and, more recently, embryo culture are proving of great value. The more basic disciplines of biochemistry and physiology, although essential to our understanding of the function of the brain, have often proved of limited value when confronted with the multiplicity of cell types, the cell–cell interactions, and the underlying crucial structural relationships that exist in the brain. The emergence of techniques that permit the perception at once of biochemical, morphological, and functional parameters has had to precede the current advances in neurobiological insight that are reflected in the chapters of this work.

The essential trace elements are not at all to be perceived as a group of functionally related substances. Each has emerged as essential because it filled a niche in evolutionary progression and the niches are not related. Thus the usefulness of copper to the organism resides mainly in the facility with which it mediates the interaction of the potentially toxic substance molecular oxygen with the organism. This is of particular importance in the brain, which is vulnerable to either a deficit or an excess of oxygen. Iodine also, through the function of the thyroid hormones, is a substance of critical importance to the brain. It is remarkable that not only these two elements, but indeed most trace metals in deficit or excess exert maximum, and sometimes specific, effects on the developing, compared with the adult, brain. A deficit of iodine during development leads to profound effects on neural development. Of the three ranges of thyroid hormone action—developmental, pituitary support, and calorigenic, the first is evident especially in the brain, with the differentiation of neurons one of the more significant targets of activity. The role of zinc in brain development is also prominent among the biological functions of that element and a deficiency during development may be teratogenic. The biochemical roles of zinc in enzymes are very numerous, for reasons that are not at once apparent, but perhaps its essential role in brain development relates to the synthesis of the nucleic acids by the zinc-containing polymerases. The normal functions of iron in oxygen transport and delivery apply also to the brain and may underlie the fairly well-recognized but perhaps nonspecific behavioral changes seen in iron deficiency. Cobalt, through the actions of its only known biologically active compound cobalamin, or Vitamin B_{12}, appears to have a specific role in maintaining the integrity of myelin, but the biochemical basis of this activity remains a matter for speculation.

Of the two remaining essential elements treated in the chapters of this work—selenium and manganese—no very specific effects of deficiency are seen in the brain. In fact, the ataxia of manganese deficiency turns out to arise from malformation of the otoliths of the inner ear. Selenium deficiency leads to reduced activity of glutathione peroxidase and may give rise to formation of lipid peroxides, but such effects in brain are not well-documented. From the range of functions mentioned for the different trace elements, it is apparent that specificity rather than conformity is the key to the evolutionary roles of these substances.

The environmental toxins reviewed are matters of growing concern and knowledge of them is rapidly expanding. The brain as a target for methyl mercury, and especially so during development has made research on the effects and mechanism of action of this substance a matter of priority. The devastating effects of Minamata disease and the tragic event in Iraq have created an acute awareness of the toxicity of methyl mercury. The nature of the changes caused by it during development are now well-documented and are described in detail in these volumes.

Manganese toxicity is also well understood. The progressive development of psychotic, neurological, and finally pathological effects are described and the evidence for involvement of specific monoamine pathways is reviewed.

Toxicity of selenium occurs in cattle in seleniferous regions, but less is known of the neurotoxic consequences than for some other elements. Selenium is remarkable in the rather narrow range of the safe intake level and both deficiency and toxicity are seen under natural conditions.

One of the most widespread and insidious of the industrial-age toxins is lead and two chapters of our work deal with this important topic. It is still not clear that defined safe levels of intake are indeed safe, because no longer is it possible to select a human population with intakes at levels as low as those before lead mining began. Considerable research effort has gone into the neurotoxic effects of lead, including attempts to analyze early behavioral changes. Once again susceptibility is much higher in the young than in later life.

The chief interest in the neurotoxicity of cadmium lies in its classical teratogenic activity exerted during development and closure of the neural tube and exhibiting the specific time-linked series of effects found with such toxins. For this reason the effects of cadmium provide an excellent field of study and a sophisticated analysis of its effects is possible.

It is of some interest that in several cases the cerebellar Purkinje cell emerges as a specific target of either deficiency of a trace element or the effects of a toxin. Abnormalities of dendritic development are found both in thyroid hormone deficiency and in the copper-deficiency of the brain that arises in Menkes' disease. Somewhat similar effects may be induced in the mouse by poisoning with methyl mercury and in developmental lead toxicity.

Distorted morphological development, gross or microscopic, has been one focus of attention in fetal and neonatal studies of trace metal imbalance. The effects are only rarely explicable in terms of biochemical mechanisms and are frequently pursued to no greater depth than descriptive morphology. It is in this area that the new techniques of immunocytochemistry may prove of value, since it links morphology directly to a chemical consitutent. One exception to the current lack of biochemical background of morphological change lies in the study of cell division by the techniques of autoradiography, and measurement of the synthesis of DNA, particularly in zinc deficiency. Biochemical analysis of the differentiation of neurons and glial cells has also been carried out in depth in the case of thyroid hormone deprivation.

Developmental studies have proved rewarding in the analysis of the role of trace elements possibly because relatively small effects in early embryogenesis may show ramified consequences. Thus much valuable information has accrued from developmental studies concerning deficiencies of zinc, copper, and iodine and from the toxicological effects of cadmium, lead, and mercury. Another possible reason for susceptibility of the immature organism may lie in the ineffectiveness of the blood–brain barrier at such times. The emergence of embryo culture as a tool, especially for neurotoxicology, permits even more effective application of agents to developing neural structures.

Neurophysiological techniques have achieved less prominence as investigational tools for study of trace elements in neurobiology, but such work is described here in relation to deficiencies of iron and zinc and for analysis of the toxic consequences of lead and mercury. As mentioned previously, behavioral measures provide very sensitive monitors for early changes in brain function and several chapters deal specifically with such measurements.

One of the primary purposes of the study of trace elements and almost the only reason for the study of neurotoxicology is the existence of recognized human disease in which the element is implicated.

By far the most prevalent human disease resulting from deprivation of a trace element is that of cretinism associated with iodine deficiency. The effects of iodine deficiency on brain development have been the subject of extensive research and it is remarkable that the reasons for the occurrence of two sharply distinguished forms of cretinism—neurological and myxedematous—are still not understood. Study of copper deficiency has been stimulated by the recognition that the invariably fatal Menkes' disease in human infants is a result of abnormal copper metabolism. Use of the mottled mouse mutant as a model of this disease has met with considerable success. The classical genetically linked human disease of zinc malabsorption—acrodermatitis enteropathia—has focused attention on the roles of zinc, but so far no genetically linked animal model has emerged. Until recently too, there was no suitable animal model for the demyelinating condition associated with pernicious anemia, but recent studies of the Egyptian fruit bat have provided evidence of comparable changes in that species and in Chapter 9, D. P. Agamanolis and his colleagues describe the experimental production of a similar condition in a primate.

As mentioned before, study of the neurotoxic affects of metals has arisen largely in response to diseases caused by environmental pollutants. Such studies are described for manganese in which the startling "manganese madness" occurs as an early sign, for mercury stimulated by Minamata disease in Japan, and by the episode in 1971–1972 that involved 459 mercury-related deaths in Iraq. Similarly, work done on both lead and cadmium has arisen from perceived and identified human toxicity-related disease.

The last chapter of the work—on the effects of lithium on manic depressives—provides yet another insight into the relationship between brain performance and the ionic environment. Lithium is not normally a significant component of the cellular milieu, but its administration to the manic depressive ameliorates an otherwise intractable disease. The reasons for the effect are still unclear, but the observation provides a stimulus and focal point for relevant research.

Index